THE BIOLOGY OF KINDNESS

THE BIOLOGY
OF KINDNESS

SIX DAILY CHOICES FOR HEALTH, WELL-BEING,
AND LONGEVITY

IMMACULATA DE VIVO AND DANIEL LUMERA

TRANSLATED BY FABIO DE VIVO

The MIT Press
Cambridge, Massachusetts
London, England

Originally published as *Biologia della gentilezza*, © 2020 Giovanni Andrea Pinna and Immaculata De Vivo

Published by arrangement with The Italian Literary Agency

The MIT Press would like to thank the anonymous peer reviewers who provided comments on drafts of this book. The generous work of academic experts is essential for establishing the authority and quality of our publications. We acknowledge with gratitude the contributions of these otherwise uncredited readers.

This book was set in Stone Serif and Stone Sans by Jen Jackowitz. Printed and bound in the United States of America.

Library of Congress Cataloging-in-Publication Data

Names: De Vivo, Immaculata, author. | Lumera, Daniel, author. | De Vivo, Fabio, translator.
Title: The biology of kindness : six daily choices for health, well-being, and longevity / Immaculata De Vivo and Daniel Lumera ; translated by Fabio De Vivo.
Other titles: Biologia della gentilezza. English
Description: Cambridge, Massachusetts : The MIT Press, 2024. | Originally published: Biologia della gentilezza: Le 6 scelte quotidiane per salute benessere e longevità, 2020. | Includes bibliographical references.
Identifiers: LCCN 2023013904 (print) | LCCN 2023013905 (ebook) | ISBN 9780262547659 | ISBN 9780262378185 (epub) | ISBN 9780262378192 (pdf)
Subjects: LCSH: Self-care, Health—Popular works. | Longevity—Popular works. | Well-being—Popular works. | Mind and body—Popular works.
Classification: LCC RA776.95 .D4813 2024 (print) | LCC RA776.95 (ebook) | DDC 613—dc23/eng/20231103
LC record available at https://lccn.loc.gov/2023013904
LC ebook record available at https://lccn.loc.gov/2023013905

10   9   8   7   6   5   4   3   2   1

To Samantha. You inspire me every day.
Immaculata De Vivo

To my mother, for the gift of her unconditional love.
Daniel Lumera

May Your rays illuminate my path.
May Your light make my mind clear.
May Your warmth heat my heart up.
May Your presence remind me of who I am.
I am. I am Light, Love, Life.
This is the time to find a new justice, that of your heart.
This is the time to build a new city, made of light, love and life.
This is the time to listen to those who cry out and are not heard.
This is the time to give, as it was given to you.
This is the time to love and to be loved, but above all,
this is the time to be love, so that every moment
still granted may be a time of peace, awareness, compassion,
brotherhood and love.
Rise, child of the Light.
Bring this light into the world. Light of the same Light.
Voice of the same Voice.
One in the One.
One in Peace.
One in the Light.

—Daniel Lumera

CONTENTS

TO THE READER

This book results from the encounter of two internationally renowned researchers, Immaculata De Vivo and Daniel Lumera, who compare science and consciousness in a revolutionary approach to health, well-being, quality of life, and awareness.

In the opening chapter on kindness, the two authors take turns discussing the "science" of attributes. A sort of biology of values is born, in which kindness is the common thread, central to their discussions throughout the book.

The book is divided into two parts. The first part is dedicated to five attributes: five key principles that thanks to the most recent scientific research and in accordance with the wisdom of ancient philosophies, we know to be essential factors for the survival and well-being of humankind on this planet. The second is dedicated to six instruments: six key strategies to cultivate and develop these values, so as to be able to live a long, healthy, and happy life. These tools can be applied effectively and concretely in our everyday lives. The last section of part II features guidelines and advice for practical ways to immerse oneself in the experience of kindness.

This is a journey into a new paradigm of health and awareness, where the rigorous language of science legitimates and explains the core values of human experience. It is a bridge between the United States and Italy, the countries in which, respectively, the authors live. The reader will also find small personal stories that step by step will turn into everyone's stories in a harmony of shared human experience.

Enjoy the reading.

For readers of this book, there is a dedicated online space where you can delve deeper into the content of *The Biology of Kindness* through videos, audio clips, and articles made available free of charge by the authors. The QR code below gives you access to this free website where you will find guided insights specific to each section, thus moving on to a practical understanding of what you are about to read and gaining inner perceptions and experiences that will make this journey even more powerful and transformative. You can access all of this additional material by scanning the QR code below or by going to www.biologyofkindness.com.

PREFACE

My love for science started in school, when I encountered the work of Charles Darwin. Scientist, explorer, and founder of the theory of evolution, his thought and work have shaped the way I see reality, teaching me the rigors of observation, perseverance in seeking answers, and need for prudence in reaching definitive or generalizable conclusions. He also taught me the value of change, the chance that we all have to improve ourselves, to adapt to an environment in constant transformation. One of the greatest wrongs we have done to this great scientist is to misinterpret the ultimate meaning of his theory, believing that natural selection favors the strongest or most intelligent. This is not true. Rather, when we study Darwin's work, it becomes clear that the most advantaged is the one who can change and adapt, making the most of the available resources.

This concept guided me through my years of university and academic training, and I found confirmation of it at every step: in all of my studies, the results of laboratory tests, and discussions with colleagues. The biological value of diversity and the ability to adapt have been the keys to human survival, the means by which we have faced the past ages, gradually evolving to become more and more skilled and intelligent.

In my years at Columbia University, Stanford, and finally Harvard, I have seen growing and convincing evidence of the significance of environmental factors, behaviors, and lifestyles for our health. The deeper we learned about the mysteries of DNA, the more clearly we understood that genetics only partially determines our well-being, and there is a lot we can do to protect ourselves from disease and extend our life. The choices we make can affect DNA, change the structure of the brain, and alter biology—leading us toward health or disease, shortening or lengthening our life, and affecting its quality. This information is too important to be left closed in laboratories and universities.

As a scientist, I felt the need to spread to the widest audience possible the incredible results gained by research in recent years and lessons that we can draw from this knowledge. If felt the need to offer a strong and clear message that invites us to play an active role in protecting our health, and make choices consistent with our view of ourselves and our vision for the future.

Five years ago, this book could not have been written because we didn't have enough evidence to measure the impact of lifestyles on DNA and therefore on an individual's overall well-being. In this short period of time, incredible progress has been made, and we now have an impressive amount of proof showing that in many cases, health can be a matter of choice and lifestyle—modifiable factors that we can change to help us stay healthy, ward off disease, and live longer and healthier lives.

Immaculata De Vivo

It was 1994 when I seriously took the path of meditation. There were many prejudices then. I was only nineteen, but my choice was radical and revolutionized my lifestyle on all fronts: eating habits, social relationships, and ways of thinking and acting. I had no idea that a few decades later science would recognize that those habits were the pillars of well-being, health, and quality of life. During my life I have never experienced anything that can come close to the sense of wholeness, completeness, and awareness that one can experience in the meditative state. I immediately understood that this was a revolutionary key to understanding not only the personal sphere but also the relational and social one.

For me, there has never been a separation between science and consciousness, spirit and matter, mind and body, and I have always felt a strong push toward a transversal approach, which has led me from a degree in natural sciences to a specialization in sociobiology, up to integrating my professional path with tools of inner research gained through meditation. Today, science is starting to confirm and demonstrate the most important philosophical and spiritual truths that the ancient wisdom traditions reached several millennia ago. This is why I felt the need to give everybody access to this knowledge about the nature of our mind along with its impact on our body and health, the process of rebalancing and healing, and quality of our lives.

At the beginning of this journey, I did not imagine that I would have the privilege of delving into themes such as forgiveness and meditation, and bringing them to universities, schools, hospitals, prisons, refugee centers, and hospices. But it could not have been otherwise because the deeper the inner search goes, the more our sense of identity expands to include every living being and gives rise to a new sense of responsibility. I also want to restore meditation to its original dimension, without diminishing it as a fad or limiting its importance to its positive side effects on physical and mental health. People approach meditation for many reasons: health, well-being, and curiosity, or to try a new experience, but there are those who feel a deep call toward research and self-realization too. This is the real key. An evolutionary key for the very survival of humankind, which leads to an authentic inner revolution, capable of restoring meaning and purpose to our lives. In this book, I wanted to share an approach to life that combines science and consciousness.

At a time when life seems to be ruled by the desire to get "everything and immediately," I would like to briefly praise discipline, which many associate with imposition, rigor, and renunciation. On the contrary, discipline is devotion, dedication, and passion. It's expressing a love that gives itself without reservation.

For twenty-six years, every day, as if for the first time, I strip naked before life, cross my legs, and close my eyes. This is how I have seen the most beautiful landscapes of this earth: the inner ones. In rain or good weather, mourning or joy, success or failure. Regardless of the fluctuations of life, I sit down and celebrate existence through silence and listening.

And this is what I wish for each of us: to learn to listen. To listen to each other and ourselves, and have the courage to follow our inner voice.

Daniel Lumera

I ATTRIBUTES

1 KINDNESS

DANIEL LUMERA AND IMMACULATA DE VIVO

Wherever there is a human being there is an opportunity for a kindness.

—Seneca, quoted in *The Cyclopaedia of Practical Quotations*, by Jehiel Keeler Hoyt and Anna Lydia Ward

THE BIOLOGY OF KINDNESS

Nowadays, science can give an exact biological correspondence to optimism, kindness, forgiveness, gratitude, and happiness, showing how fundamental these values are for living long, healthy, and happy lives, but above all for the survival and evolution of humankind on this planet. The encounter between science and conscience gives birth to the idea and experience of the *biology of kindness*, a "biology of values" that demonstrates how changes in terms of awareness in one's inner world can positively affect biological parameters, DNA, well-being, health, and longevity as well as the quality of relationships and social processes.

The biology of kindness explores the biological, vital, emotional, mental, social, and spiritual impact of five values: kindness, optimism, forgiveness, gratitude, and happiness. It also provides tools and strategies for health and longevity through the six pillars of well-being: happy relationships, nutrition, meditation, physical activity, music, and contact with nature.

This approach considers the multidimensional nature of the human being, in which interdependent levels such as body, emotions, mind, consciousness, relationships, and nature mutually influence each other, impacting the individual, relational and social well-being, and that of the entire planet in a closely interconnected system.

The biology of values wants to show through data and experiences that it is not the physically, mentally, and economically strongest but instead the kindest one that is be best suited to change and survive on this planet. Kindness turns out to be the best evolutionary strategy for living long, healthily, and happily. Cultivating and developing the five values proposed in this book is therefore no longer just a moral, ethical, or social question but also an essential evolutionary must.

The biology of kindness is a journey into understanding the power of the mind over genes, secrets of longevity, anti-inflammatory and antiaging processes achieved through meditation, relationship between nutrition and cancer, impact of nature and music on health and mood, and importance of knowing how to create happy relationships for health and quality of life. It's a bridge that unites the knowledge of ancient millenary traditions with modern scientific evidence, showing the new frontiers of health and well-being.

BEING KIND

Being kind to ourselves and others is, as we will see, not only a biological matter of health, well-being, and longevity but also an evolutionary strategy useful for the entire human race to survive.

The word *kindness* (*gentilezza* in Italian) recalls a sense of politeness, sweetness, and lovableness. In Latin, *gentilis* derives from *gens* and indicates an extended family group, a clan to which one belongs.

The *gens* in ancient Rome was a sort of extended noble family with mutual duties of defense and assistance, and lacking close relatives, gave the right of inheritance. Even a burial site was shared among the members.

Gentilem in Latin means "belonging to the *gens*"—that is, an aristocratic family—a social condition to which corresponded moral and aptitudinal qualities such as real courtesy, politeness, and grace. They were not just formal external behaviors but rather real inner feelings: the nobility of the soul that expresses high qualities.

Kindness goes far beyond the common meaning of good manners. It is a social value of fundamental importance that shapes a sense of belonging with no need to resort to violent verbal communication, create competition or make enemies, or leverage primary instincts, fears, and emotional wounds. Belonging to *gente* ("people" in Italian) is an inclusive process whose characteristic elements are empathy, courtesy, kindness, and spirit of service.

Being *gentile* thus requires and implies still today a nobility of mind capable of expressing that sense of belonging based on mutual recognition, respect, and benevolent care.

Kindness, as an indispensable and essential social principle, should be the basis of any relationship between human beings so that they can relate in the most useful, fraternal, and elevated way possible.

The seed of genuine kindness, like the lotus flower, has the power to grow and blossom even in the mud.

True change always begins with small gestures: nothing is more immense than the tiny seed of a kind thought. A kind thought that with its evolutionary power, drop by drop, digs even the hardest of rocks: hatred.

So let's not deprive ourselves of the satisfaction of responding with kindness to fear, rudeness, revenge, abuse, ignorance, violence, or resentment. Also for reasons of health and quality of life.

In the intimacy of our feelings, we could start by being kind to ourselves.

Kindness everywhere. Even in silence. Among the notes of existence.

KINDNESS IN THE THREE REALMS

Let's get used to practicing three conscious acts of kindness every day.

- The first toward a human being
- The second toward an animal
- The third toward a plant

PASS THE FAVOR: BENJAMIN FRANKLIN AND TEN GOLD COINS

It was only a sunny smile,
And little it cost in the giving;
But it scattered the night
Like morning light,
And made the day worth living.

—Anon, "Only," in *The Western Teacher*, March 1893

(continued)

In 1784, Franklin received a letter from an old friend, Benjamin Webb, asking for financial help. Having limited means, Franklin was reluctant to agree, but eventually did, on one condition: his friend should pay off the debt not by returning the money to him but rather by donating it to someone else with similar needs. Franklin thought that his investment would help not just one person but instead more and more people in an endless chain of good deeds.

So along with ten golden coins, he wrote this letter to his friend:

> I send you herewith a Bill for Ten Louis d'ors. I do not pretend to *give* such a Sum. I only *lend* it to you. When you shall return to your Country with a good Character, you cannot fail of getting into some Business that will in time enable you to pay all your Debts: In that Case, when you meet with another honest Man in similar Distress, you must pay me by lending this Sum to him; enjoyning [*sic*] him to discharge the Debt by a like operation when he shall be able and shall meet with such another opportunity.—I hope it may thus go thro' many hands before it meets with a Knave that will stop its Progress. This is a Trick of mine for doing a deal of good with a little money.

Franklin could not imagine that no obstacle could have stopped the wave generated by this small act of kindness, from which an entire movement would be born, a movement better known today as "pay it forward" or "pass the favor."

The concept of pay it forward coined by Franklin has ancient origins: Ralph Waldo Emerson wrote about it, science fiction masters Robert Heinlein and Ray Bradbury illustrated it in their work, and some believe that it even dates back to the Greek playwright Menander in 300 BC. It has expanded by taking root in today's culture, becoming first a book, then a film, and then causing a worldwide cascade of endless kindness. April 28 is International Pay It Forward Day, a celebration of altruism and the good of humanity that in 2019, saw people from eighty-six different countries unite in hundreds of millions of small acts of kindness.

How does it work? It's simple: you just do a little act of kindness without expecting anything in return. If you receive it, you do it in turn for someone else and keep the ripple effect alive. Here are some ideas:

- Help an elderly woman carry shopping bags
- Offer a homeless person a meal
- Donate blood
- Give a hug
- Leave a paid fruit and vegetable basket
- Offer a ride to an acquaintance
- Make a donation to a social project
- Leave a note of appreciation for a colleague
- Donate your time and talents to someone who needs it

- Plant a tree
- Take a dog from the shelter for a walk
- Share your action on social media and spread the wave of kindness
- Imagine doing this simple practice in every moment of your life. What would happen if we all did one small gesture of kindness each day?

DOING ONESELF GOOD BY DOING GOOD TO OTHERS

Everything we do in a selfless way, without profit and only aiming at making somebody else feel good, is kindness. It can have various shades, to which we give different names. Sometimes it is altruism, sometimes compassion, and often it is empathy, gratitude, and generosity. All different forms of the same feeling of love toward others, which pushes us to perform actions for the sole pleasure of doing it, without asking for anything in return. We all do it, even when we feel we are too busy with our own needs to pay attention to those of others. We do it in small doses, in negligible events of our day, yet more frequently than it seems to ourselves. And every time there is someone who smiles for our generosity, someone to whom we have done good or transmitted a positive emotion without expecting any counterpart.

It is a profound trait of human beings, useful for evolution because it favors the creation of social bonds and encourages collaboration, making individuals more willing to give up a piece of their selfishness to build something together with others. From small gestures to large solidarity initiatives, kindness is the best way we have to connect with others, communicate, solve problems, and achieve goals. And it is a broad concept, as we said, that includes many nuances but starts from the same spontaneous desire of wanting the good for others, both for the people we know and strangers, out of an instinctive impulse that comes from our deepest humanity.

The whole set of values and tools that we present in this book as strategies to protect health and promote longevity can be contained within the concept of kindness, or rather that of the biology of kindness. Because it is a wealth of practical resources that calls into question our deep feelings, as elusive as emotions are, but that can at the same time satisfy my need, as a scientist, for tangible evidence, solid and clinically relevant data, and credible numbers and percentages. Kindness, with all of its forms, proves itself scientifically valid as a preventive tool, support to therapies, and way to physical and mental health.

KINDNESS IS HEALTH

The fact that kindness and in general the positive feelings of humanity and compassion play a role in improving health has been widespread in the scientific world for many years, but only in recent times has this knowledge been actively used, with meaningful and encouraging results.

The most advanced cancer research centers in the world have been offering psychological support protocols for patients and their families focused precisely on kindness as a means of human closeness to the people who face the disease. It's been seen over time that kindness is a powerful tool that can defuse the negative emotions associated with cancer diagnoses and therapies, helping in some cases to improve the response to treatments. Based on a long experience in researching and treating this disease, scientists from different institutes have outlined six ways of using kindness in the treatment of cancer, activating protocols that involve patients, families, and health care professionals.

SIX ACTS OF KINDNESS AGAINST CANCER

1. Listening—Health care professionals take time to deeply understand the needs and concerns of the patient and their family

2. Empathy—Doctors and nurses establish a strong, understanding bond with the patient and outline the treatments, trying to prevent any avoidable suffering

3. Generosity—The operators perform acts of kindness and care that go beyond the expectations of the patient and families

4. Antistress—Targeted assistance and caregiving practices to reduce stress and anxiety in the patient

5. Honesty—Always inform the patient about the state of the disease and the therapies, using the right words and conveying positive emotions

6. Support—Supporting the role of families, whose mental and physical well-being is essential for the effectiveness of patient treatments

---

Kindness-based interventions have also been successfully tested in cardiovascular disease. Emotions, triggering stress mechanisms that affect heartbeat and blood pressure, have a strong correlation with this category of illnesses, and the possibility of preventing them or improving the conditions of those who already suffer from them through the use of positive feelings has been an important scientific insight. Over the years, the confirmations have multiplied, giving us today the opportunity to consider kindness, together with gratitude, altruism, and empathy, as tools for defending our health.

A 2011 study by Harvard University observed the effects of a "positive psychology" intervention on patients hospitalized for cardiovascular disease, particularly acute coronary syndrome and heart failure. The eight-week intervention protocol included three categories of exercises based on kindness, optimism, and gratitude. At the end of the experiment, there were signs of improvement in the clinical conditions of the subjects observed, despite the intervention being short and their disease being severe.

With this in mind, further studies have been conducted to verify the chance of using kindness in the prevention of cardiovascular diseases and not only as a support for treatment after the onset of a disease. A team of researchers from various US universities has analyzed this possible link in a population composed of Hispanic Americans, statistically subject to a higher risk of cardiovascular events than individuals of European descent and therefore considered in need of a targeted intervention program. In particular, hypertension was observed as a risk factor, related to the possible effects of a positive psychology intervention. The researchers activated an eight-week protocol of interventions through therapists and social workers for an average of 90–120 minutes a week, observing how the blood pressure data changed, but also other indicators such as emotional well-being, psychological serenity, likelihood of healthy behaviors, and inflammation markers. By the end of the program, which included exercises such as reconsidering stressful events, performing acts of kindness, and expressing gratitude, hypertensive individuals had lower blood pressure levels and responded positively to various indicators of psychological and emotional well-being.

Techniques based on acts of kindness have also been used to support behavioral therapies for people with social phobia, a type of anxiety that prevents them from establishing conventional social relationships or facing certain types of contexts in which interaction with others is necessary. A 2015 Canadian study analyzed a population of 146 college students suffering from the phobia who filled out a questionnaire to measure the level of social anxiety experienced. The questions concerned cognitive (for example, "I'm worried about expressing myself for fear of appearing embarrassing"), emotional ("It makes me nervous to deal with people I don't know well"), and behavioral aspects ("I have difficulty maintaining eye contact with others") to establish the baseline for each person. Participants were asked to practice at least three acts of kindness two days a week for a period of four weeks, defining acts of kindness as actions performed for the benefit of someone else with no benefit to themselves, but instead assuming a cost. Some instances of the deeds performed by the study participants were preparing dinner for a roommate, mowing a neighbor's lawn, or donating a sum to charity. The subjects were then exposed to social situations that cause anxiety, up to three a day for two days a week for four weeks.

Looking at all the data collected, the researchers concluded that practicing acts of kindness can significantly reduce the levels of social phobia—more specifically by reducing the occasions when the person avoids a situation for fear of experiencing anxiety—and is a phenomenon that persists over time. Levels of perceived anxiety are reduced too, and this result, thanks to the role played by kindness, is achieved at a significantly faster rate than the exposure-only technique, in which the anxious subject is invited to immerse themselves in a situation that puts them in discomfort and endure the peak of anxiety until it naturally subsides. The positive reactions received as a result of an act of kindness push the person to no longer feel with the same intensity the need to avoid social situations or expect only negative consequences from interacting with others. Focusing on the good of someone else in a selfless way therefore seems to have a strong impact on our emotional balance mechanisms, making it a powerful tool for well-being, able to improve the relationship with others and our own health.

KINDNESS IS HAPPINESS

In 2018, the University of Oxford in the United Kingdom observed a sample of 683 individuals to investigate the effect of kindness on overall happiness. Specifically, the research wanted to compare the consequences of acts of kindness practiced toward friends with those carried out toward strangers. The subjects practiced kindness every day for a week, and afterward, their happiness levels significantly increased. The remarkable aspect is that there was no difference based on the intensity of the bonds, so acts of kindness performed toward friends or strangers cause the same effects on happiness. Kindness is therefore a positive act in itself, which is good for us regardless of the identity of the people to whom we dedicate it.

THE CHILD AND THE STARFISH

Once upon a time there was a man who lived near a beach. Every day he woke up and started off by taking a walk on the sand. One day, after a terrible storm, he found thousands of starfish writhing in agony on the seashore.

The man felt bad about the situation. He knew that starfish could not live more than five minutes out of water. All of those creatures would be dead in a short time. "How sad!" he thought. No idea came to him, however. The phenomenon had gathered a crowd of people on the beach. Everyone was watching, and nobody did anything. After a while he saw a child leave their parent's hand, take off their shoes and socks with great enthusiasm, and run on the beach, picking up the starfish one by one and throwing them back into the water. The child was all agitated and sweaty. "What are you doing?" the man asked him. "I'm returning the stars to the sea," replied the child, who was clearly already fatigued. The man stopped for a moment to think. What the child was doing seemed absurd to him and he couldn't help saying what he thought: "But it's useless! There are thousands of starfish on the shore, you'll never make it!" The child, who was holding a starfish in their hand, smiled at him, bent down to pick up another one, and throwing it into the sea replied, "For this starfish, yes, it makes sense!" The man remained silent for a moment, then took off his shoes and ran to the seaside to help the child. A moment later a couple of other children arrived. In a few minutes there were ten, one hundred, two hundred, and then thousands of barefoot people, by the sea, throwing the starfish back into the water.

KINDNESS IS WONDER

Violence comes from our soul. Small drops of violence fall every day. One after the other. In a world where everywhere there is something wonderful to observe, we are distracted. Distracted by living in a hurry, too full of stimuli, work, noise, television, news, social networks, meetings, people, and thoughts. Always doing something. And often doing it fast without stopping. Isn't this violence? Isn't this the soil in which the seed of violence grows? Yet it would take so little. A little wonder would be enough. Because we are surrounded by a constant miracle. Only a few people take the time to be quiet and listen. Listen and feel. Allow oneself the privilege of finding free spaces to look at the flowers. The miracle of flowers. Kindness can stop the world. And make us breathe. And remember. That flowers still grow everywhere.

We should look at the world and life free from our knowledge. Without judgments, prejudices, concepts, ideas, thoughts, and desires. Becoming a child again, and looking with clear and pure eyes, with the same wonder as when the little ones see things for the first time. Have you ever seen how children react when they first discover the rain or the existence of their shadow? What a privilege to live with that sense of wonder forever. Wonder and kindness. Just these two things would be enough in life.

Being a child at all times is a choice that must be repeated constantly because our eyes, in every moment, collect dust. We must always remember to clean them. We cannot expect them to remain pure if we do nothing. The simple passing of time makes them dirty and dull, and not because we have done something wrong. But closing a house so that it doesn't get dirty is not enough. The dust will settle anyway. Let's remember to always clean our eyes, with kindness, to look at the world with the purity of a child.

2 TELOMERES

IMMACULATA DE VIVO

We are what we repeatedly do.
—Will Durant, *The Story of Philosophy*

THE SENTINELS OF LONGEVITY

Being kind, practicing altruism, and living peacefully with others are all habits that make us feel better. So we are told by common sense as well as the anonymous proverb: "Practice random kindness and senseless acts of beauty." But there is also a scientific basis for this. In recent decades, science has examined every cell in our bodies under the microscope looking for confirmation of what we have instinctively always known: that living in harmony with ourselves and others can give us a long, serene, and healthy life.

Today we have enough evidence to say that our lifestyle matters much more than genetics in determining our overall health. Chronic diseases such as diabetes, cancer, and heart disease often have roots in a genetic predisposition, but this alone is rarely sufficient to cause the onset of a disease. There are also environmental factors, the substances to which we are exposed, and the life choices that we make. "Genes load the gun, the environment pulls the trigger," we scientists say. A loaded gun can't hurt anyone if there's no hand to pull the trigger. That hand is our own. We can decide to stay away from the weapon. Or we can decide otherwise, thus "helping" genetics to harm us. The choice is ours.

But how do we know what is good for us and what is bad? Certain molecules in our body can tell us whether we are predisposed to a certain disease. These molecules are called *biomarkers* and act as "sentinels" of our health. Their presence or absence, concentration, and biological character-istics can tell us a great deal about how likely we are to develop pathological conditions. Among the many biomarkers known to science, one group in particular has proven especially useful in recent years, giving us informa-tion about our health and life expectancy. Telomeres, today's "superstars," have captured the imagination not only of scientists but of the general public too.

WHAT ARE TELOMERES?

Telomeres are DNA structures at the ends of chromosomes that protect chromosomes from damage and keep the genetic material of a cell intact. Cells in our body reproduce all the time to replace those cells reaching the end of their life cycle. In this replication process, telomeres lose small segments of their genetic strand, so the new cell will have slightly shorter telomeres than the one from which it was generated. This is a completely natural fact, which by itself has no significant consequences. All of us are born with a certain telomere length, which will progressively shorten over the course of our lives. This process is irreversible; lost telomeric material cannot be recovered, nor can the original telomere length be restored. As they get shorter, telomeres struggle to perform their protective function. Once they reach a critical length, the cell stops replicating and undergoes a programmed death process. For this reason, science considers the length of telomeres a real biological clock that determines the life span of a cell, and by extension, the organism to which it belongs; longer telomeres are associated with long-lived individuals, while shorter telomeres are associ-ated with a shorter life expectancy.

TELOMERES AND ENVIRONMENTAL FACTORS

Telomere shortening is due not only to a natural process but also influ-enced by environmental factors and lifestyles. Smoking, alcohol abuse, a sedentary lifestyle, a poor diet, and obesity contribute to accelerating and aggravating the telomere shortening, and therefore cell death, creating the conditions for the onset of disease and premature aging of the body. An individual's psychological condition can also have a significant impact on their cellular health. Stress has been recognized as one of the greatest enemies of our telomeres because it involves an oxidative process and state of inflammation that damages telomeres and promotes their shortening. We cannot eliminate the causes of stress because they are frequently outside our control, but we can change the way we respond to stress, adopting protective strategies, such as playing sports, helping others, or walking in nature—in short, practicing any good habit that helps us to relax and defuses the stress pathway.

Healthy lifestyles based on proper nutrition, moderate physical activity, and no smoking have proven to protect telomeres from excessive wear, thus promoting overall health and longevity. Alongside these "universal recipes," which are valid for everyone, there are other strategies that each of us can adopt individually. For one person a certain strategy will be highly effective, while for someone else, a different strategy will be better. Let's imagine that we want to grow plants. Seeds that have good soil and enough water will sprout, while seeds planted in less favorable conditions will wither and die.

TELOMERE REVOLUTION

The revolutionary discovery of telomeres and their mechanisms lies precisely here: we have learned that our DNA is modifiable and that our life choices can transform our genetics. Our DNA is not immutable, but it does respond to the stress of the external environment, adapt to continually evolving conditions, and express positive or negative potential depending on what inputs we provide.

This explains why scientists decided to use telomeres as indicators of an individual's overall health, particularly their susceptibility to chronic diseases such as diabetes, cardiovascular issues, tumors, and osteoarthritis. These telomeres are called biomarkers because they provide us with valuable information about our biological condition. Science has identified many biomarkers, able to convey a variety of information, but telomeres have proven to be especially interesting due to their ability to adapt to environmental changes. Observing telomere length in individuals living in varying contexts, with certain types of stress, gives us invaluable information about what favors good health as well as what determines disease, premature aging, and early death.

Research in recent years has gone even further. After confirming the health benefits of a plant-based diet, moderate physical activity, a regular sleep cycle, and other habits known to be beneficial, scientists began to investigate the influence that other behaviors and attitudes may have on our genetics. Meditation is known to reduce stress, but such positive attitudes as kindness, optimism, empathy, openness, and compassion also help with stress reduction, and thus are considered "friends" of our telomeres—which in fact are longer in people who adopt these types of behaviors. This is an extraordinary line of investigation that has given us, for the first time, scientific proof that living in harmony with oneself and others means delayed aging, avoiding disease, and living longer and healthier lives. A new biology of kindness that can help us using simple but impactful practices that are good for us, the people around us, and our planet.

DNA, THE BOOK OF LIFE

DNA is a molecule that contains all the genetic information that determines our biology. It's a double strand, spiraling around itself, forming clusters called chromosomes. Each one of us has twenty-three pairs of chromosomes, contained in the nucleus of each cell in our body. The basic units of DNA, called nucleotides, are of four different types, each indicated by a letter: A (adenine), G (guanine), C (cytosine), and T (thymine). RNA, the molecule that allows DNA to translate information into proteins, has uracil (U) instead of thymine.

Just as the English alphabet is made up of twenty-six letters, so the DNA alphabet is made up of only four letters, which are combined in a great variety of different sequences. Each sequence is like a word in the natural language, capable of expressing precise information; this sequence is called a gene. If in English the sequence of letters C-A-T gives us the word *cat*, in the DNA language the AUG sequence means "beginning"—that is, the start of a protein synthesis process, while the UGA sequence means "end" because it interrupts the biochemical reaction in progress.

DNA is therefore expressed in a real language, through words of three letters, each concatenated to form sentences and then entire paragraphs of increasingly complex information to form our genome—that is, in the end, the entire book of our life. It contains all the genetic information that makes us who we are, each one of us different and unique.

TELOMERES AND AGING

In 2009, Elizabeth Blackburn, Carol Greider, and Jack Szostak were awarded the Nobel Prize in Medicine for their discoveries related to the functioning of telomeres and the telomerase enzyme.

Since then, these almost-unknown structures of DNA have become celebrities even outside the scientific world, thanks to the suggestive possibilities they have opened up in research against cancer and other chronic diseases. The protective mechanisms of telomeres and their progressive shortening have been long known in the academic world, but the further studies by the Blackburn team and recognition of the Academy of Sweden have brought telomere science out of the laboratories and into the public arena. It's a bit like when a movie wins an Oscar: everyone runs to the theater to see it, talks about it at the bar, and reads about it in the newspapers. From a subject for specialists, telomeres quickly became a topic of conversation, which in turn gave rise to several new studies that are going on now and will hopefully continue in the future.

The first thing to know about telomeres is that they are the most important and significant markers of aging, since their shortening is naturally connected with the passage of time. At the time of birth, each of us has a genetic heritage of chromosomes and the telomeres that protect them. About 60 percent of their original length is determined by heredity. This is our initial capital, our biological "entrance fee." From that moment on, through the natural process of cellular replication, telomeres will progressively shorten, and over time this will reduce their ability to protect our DNA from various types of attacks. If we measure the telomeres of a newborn baby and compare them with those of an elderly person, we will notice a reduction in their length, which has occurred in a natural way over the years.

This reduction is an indicator of cellular senescence, the scientifically visible sign that cells, tissues, and ultimately the entire organism are aging. There is a building in Boston called the Old John Hancock Building, which is topped by a tall, illuminated flagpole. Since 1950, this lightning rod lights up red or blue, sometimes steady, sometimes flashing, to indicate the weather forecast. Bostonians have learned to crack the code and know what the weather will be like tomorrow simply by glancing at the city skyline.

This is roughly the function telomeres serve for scientists: their character-istics allow us to predict what our cellular health is likely to be in the near future—"what the weather will be like" in our biology if today's life condi-tions remain unchanged.

THE BAD NEWS: PREMATURE AGING

So telomeres get shorter year after year, and we can't avoid it. So why do we care? Because this process does not happen at a constant and invariable rate but instead can be influenced by environmental factors. This is the most significant discovery on the mechanism that regulates telomere func-tion: certain factors can accelerate their shortening, helping to anticipate the onset of diseases typical of aging. In addition to the normal wear and tear due to aging, which we can't control, further damage occurs due to bad habits, prolonged exposure to harmful substances, and psychological stress. In several studies, people exposed to such factors had shorter telomeres than members of the control groups. Their bodies were in a sense "older"—not in terms of chronological age, but in relation to cell and tissue health. For example, in 2017, data were collected from eighty-four different studies on the effects of cigarette smoking on telomere length. The connection between smoking and the early appearance of diseases related to aging had been clear for some time, but it wasn't apparent whether smoking was also a biological aging accelerator, of which telomeres are the most reliable indi-cators. Put simply, we know that smoking makes us sick, but does it make us age earlier as well? From extensive research, it has been determined not only that telomere length is shorter in smokers than in nonsmokers but also that telomere damage is greater in those who have smoked, even for a short period in their life, than in those who have never touched a cigarette. Exposure to tobacco smoke, then, can be considered a factor involved in premature cell aging, and in turn, a precursor of common chronic diseases. Thanks to telomeres, a person's state of health, their level of cellular aging, and the probability of developing diseases have all become scientifically measurable. Shorter telomeres result in shorter life expectancy; longer telo-meres indicate longevity. Thus we can define telomeres as "biomarkers of aging," or to put it more poetically, the sentinels of longevity.

Telomere shortening is, in fact, associated with the onset of numer-ous diseases, and more generally, a reduced life expectancy. This happens because the DNA in the cells, no longer adequately protected by telomeres, is more easily damaged. Various pathological conditions are related to telomere shortening; diabetes, cardiovascular disease, cancer, Alzheimer's disease, and dementia are some of the most common. In laboratory studies, patients with these diseases had shorter telomeres than did healthy people. We cannot say that shorter telomeres are the direct cause of these diseases, but certainly we can recognize that telomere attrition creates a favorable environment for the development of diseases related to aging. If we cut down the trees on the side of a mountain and later a landslide occurs in the aftermath of a storm, the direct cause of the disaster would be the heavy rainfall, but the deforestation will have helped create the necessary conditions for the event. In medical science, a correlation between two events is not necessarily a causal link, but it is still significant.

Also significant are the results of one of the most important studies on the link between telomere length and mortality, conducted at the University of Copenhagen in 2015 on a sample of nearly sixty-five thousand people. Starting from the premise that shorter telomeres are related to aging and diseases typical of old age, scientists asked whether they are also associated with higher mortality due to cancer or other diseases. In other words, they wondered if people with worn-out telomeres are more at risk of dying than people with longer telomeres. They studied the DNA of the participants over a few years and found that the people who died during the course of the research study had shorter telomeres than the survivors. The leading cause of death was cardiovascular disease, followed by cancer. This fundamental research has allowed us to understand that excessive wear and tear on telomeres can play a role in accelerating an individual's death from chronic disease.

THE GOOD NEWS: MODIFIABLE FACTORS

Premature aging, chronic disease, and shortened life expectancy: Do these telomeres bring us only bad news? Absolutely not. On the contrary, their susceptibility to environmental factors and lifestyle is valid in both senses: we can accelerate their attrition—by smoking, for example—but we can also limit it as much as possible, simply by adopting healthy lifestyles and antistress behaviors. Scientifically valid methods are physical activity, healthy eating habits—especially that inspired by the Mediterranean diet—abstinence from cigarette smoking, and stress reduction behaviors, which essentially lower the levels of cortisol, commonly known as the "stress hormone," in the bloodstream. Meditation, spending time in nature, listening to music, and maintaining a positive attitude are all strategies we can use to protect our genetics and preserve our health. We can't lengthen our telomeres, but we can slow down their attrition as much as possible.

This complex interaction between environment and health is at the core of an important study by Harvard University dedicated not only to telomeres but more generally to the risk factors of the most common chronic diseases. The Nurses' Health Study (NHS), begun in 1976 and still ongoing, is one of the largest epidemiological studies ever conducted, and is a true mine of information on the mechanisms of many diseases and the most effective strategies to protect our health. The great merit of this research is that it highlighted the influence of lifestyle on our bodies. Over the course of more than forty years, the NHS has studied the daily habits and health status of over two hundred thousand US nurses. In recent years, the NHS has started analyzing telomeres too, finding numerous confirmations to the hypothesis that bad lifestyles damage our health. Many will argue that these consequences are already known; we all know that smoking, eating junk food, and being sedentary is bad. And this is true. But now we have additional scientific evidence and are learning about the mechanisms through which environmental factors affect our well-being, thereby allowing us to develop new treatments and defense strategies. More important, telomere science tells us that environmental influence on the body can modify our DNA, and that bad habits or adverse life conditions can irreversibly damage our health. At the same time, it tells us that we have weapons on our side: strategies we can adopt to minimize this damage.

A key lesson in this sense comes from the study of centenarians, particularly in the so-called Blue Zones—that is, those areas of the world with a singular concentration of people who live over one hundred years in perfect health. It is evident that studying the characteristics of these people and their environment can help us understand if there is a "recipe for longevity" and what the ingredients may be. There are five Blue Zones: Sardinia (Italy), Ikaria (Greece), Okinawa (Japan), Nicoya (Costa Rica), and the Seventh Day Adventist community of Loma Linda (California, United States).

These are singular groups, characterized by a certain degree of isolation and unusual living conditions. Scientists studied the genetic profile of these people and considered their daily habits, diet, and environment. They collected an impressive amount of data looking at the distinctive elements of these communities, hoping to find an explanation to their people's extraordinary longevity.

The first factor investigated was obviously DNA, in order to understand the genetic baseline prior to environmental influence. Research has shown that telomere length in centenarians is determined by hereditary factors in a higher percentage than in people with an average life expectancy. In fact, it was found that the parents of centenarians have a longer life on average than those of noncentenarians, and that their children have longer telomeres than those of the control groups. There is therefore an underlying genetic variable on which the particular environmental conditions of these communities are grafted. From a genetic and evolutionary point of view, isolation is generally an unfavorable condition because in the presence of a new threat, such as an unknown virus or significant variation in the diet, the isolated community may not have adequate genetic defenses to survive. In this case, however, isolation played a positive role, favoring the circulation of "longevity genes" within the group and thus creating the conditions for an unusual concentration of centenarians in a relatively small geographic area.

One of the most significant environmental factors studied was nutrition. We will consider this topic in more detail later, but for now it is useful to mention the general characteristics of a diet associated with health and longevity. A common element of the Blue Zones' populations is a diet rich in antioxidants and substances that reduce inflammation—two groups of nutrients abundant in plant foods. Their action counteracts oxidative stress and inflammation—that is, the biochemical processes that accelerate the wear and tear of telomeres, thereby implementing a protective action for the DNA.

Current scientific evidence, then, tells us that aging is indeed an unavoidable process, but it is only partially determined by genetics. It is caused by a combination of hereditary predisposition, environment, and lifestyle. As such, we can take actions to slow it down, and try to make our coming years more peaceful and fulfilling.

STRESS, THE INVISIBLE ENEMY

Let's imagine we are in the savanna, a hundred thousand years ago, scanning the horizon in search of food and danger signs. We have senses on alert, sharp eyesight, and ears ready to perceive the slightest noise. From the thick vegetation we see a threatening-looking lioness emerge. Our body's reaction is immediate: our heartbeat accelerates, blood pressure rises, and blood sugar level skyrockets to supply energy to all muscles, especially those in the legs. We immediately run to safety; in one leap we try to get as far away from the danger as possible. We don't perceive pain; we don't feel any kind of need at that moment. The feeling of hunger, which earlier had prompted us to venture into the open plain in search of food, has completely disappeared. What's happening in our body is one of the most fascinating and complex defense mechanisms that evolution has endowed us with, and an explosion of rapid biochemical reactions that have only one goal: survival. To do this, the body is willing to maximize resources as much as possible, even at the cost of suspending fundamental vital functions. Digestion slows down and the reproductive system shuts down temporarily to allow the body not to waste precious energy in activities that are useless at that time. This state of altered biochemical functions is meant to last only for a short time, giving us the resources to either fight the lion or run to safety—"fight or flight." Once the danger has passed, the body gradually returns to baseline because the acute emergency response is no longer needed.

This fight-or-flight mechanism allows us to use all of our body's resources, either to face the danger or escape from it. This is one of the greatest secrets of our survival on this planet and the result of millions of years of adaptation to the environment. And it continues to be useful to us, not only in the unlikely circumstance of coming face-to-face with a hungry lioness, but also when we escape from other forms of danger, such as a fire, shark, or even Freddy Krueger. Any of these circumstances would require an immediate survival reaction, which will be intense and possibly violent, but in any case, short.

Let's imagine an apparently different context. We are sitting in the car, stuck in traffic on Interstate 405 in Los Angeles at rush hour. Frustrated, angry drivers honk their horns and cut in front of us, and we're anxiously thinking that we will be late. But meanwhile, there is the mortgage payment to be paid, the children to drive to their music lesson, and the mother-in-law coming for the weekend. This is not the lioness emerging from the grasses of the savanna—it's not a matter of life or death—but our body doesn't know this. The psychological pressure we are subjected to triggers the same type of biochemical response in our bodies: it accelerates the heartbeat, increases blood pressure, and contracts the muscles. This reaction is not supposed to last long because in a short time, we should be able to escape from the stressful situation and our vital functions should return to normal. Yet we can't do this because we have no control over the situation, so our biochemical alteration continues, perhaps for months or even years.

This continuous wear and tear, not anticipated by evolution, has detrimental effects on our health and can lead to the early onset of diseases associated with aging. This emergency state becomes our new normal. Our emergency response functions are activated—hypertension, tachycardia, and muscle contraction—while other, more useful functions—digestion and reproduction—are deactivated or attenuated. Hence what originally developed as a formidable survival mechanism, the result of an extraordinary process of evolutionary adaptation, has become an enemy that can eventually damage our health. How did we get here?

FROM ACUTE TO CHRONIC STRESS

The environment in which we live has changed enormously since those early days in the savanna. Our complex network of relationships, the psychosocial pressures we experience every day, are not remotely comparable to those of the prehistoric era. Civilization developed too quickly for our biology to adapt to it. Mechanisms refined over millions of years have made us suitable for a certain type of natural environment that has been totally transformed in a relatively "short" period of time in terms of evolution. In only three thousand years, we have gone from Uruk cuneiform tablets to smartphones, from agricultural villages to metropolises of steel and glass. The amount of stimuli we get each day has become significantly more pressing, and although we no longer risk our lives to find nourishment, our lives have become much more stressful than they were in the past. So from a purely scientific and evolutionary standpoint, all of this stress comes from ourselves, or rather from the civilization we have built.

According to Robert Sapolsky, a neuroscientist at Stanford University, the stress we experience in life depends on our high level of intelligence. In a provocative but scientifically accurate way, Sapolsky states that we humans, like other primates, suffer the effects of stress because we have "too much free time." A time we spend in activities that cause stress to ourselves and others: competing for territory and resources, fighting over petty rivalries, or continually creating states of conflict. And since chronic psychosocial stress can compromise our health, we can say that evolution has made us so intelligent that it makes us sick. Intelligent and at the mercy of our emotions: the emotional part of our mind is involved in the mechanisms of stress. This complex emotional life creates a psychological suffering not seen in other mammals.

Sapolsky has been studying the behavior of baboons and our other "close relatives" for over thirty years, monitoring their health conditions and the most common diseases. He says these primates are a good model of observation because like many humans, they are not exposed to life-threatening stressors. A herd of Serengeti baboons, for example, is not especially threatened by other predators and searches for food for about three hours a day, thus having nine hours of free time that they can spend creating stressful psychosocial situations. Their real enemy is not outside but rather inside the group: it's themselves. Analyzing the cells and tissues of these animals, Sapolsky and his team observed that individuals under stress have reproductive problems, difficulty in healing wounds, high blood pressure, and dysfunctions in the biochemical mechanism that regulates anxiety in the brain. These symptoms are also found in humans, suggesting that this combination of intelligence and emotional depth is the cause of stress, which triggers fight-or-flight mechanisms and then extends them over time, with significant health consequences.

STRESS AND DISEASE

When we suffer from stress, anxiety, and worry, our brain releases into the bloodstream several hormones, which prepare the body for the fight-or-flight response. Regulated by the hypothalamus-pituitary-adrenal axis, these reactions mainly produce adrenaline and glucocorticoids—including cortisol, the stress hormone, which quickly reaches muscles and organs and crosses the brain membrane. Cortisol has numerous physiological effects, all of which are beneficial if you need to fight an enemy or run from danger: increased blood sugar levels, hypertension, appetite reduction, heightened attention and memory, and reduced perception of pain. Relatively short episodes of stress and anxiety may not leave lasting marks in our body, but prolonged exposure to these hormones has been linked to the onset of various diseases. Traumatic events can cause latent damage, whose symptoms may occur later, sometimes even years after the traumatic event itself. More generally, prolonged stress and anxiety have been linked to chronic disease and premature aging.

The most common stress-related pathologies are cardiovascular—hypertension and heart attack—but there are also ulcers, gastrointestinal and renal dysfunction, and various psychological illnesses. A 2013 Australian research study identified stress from work or family as a probable cause of type 2 diabetes, cancer, circulation problems, asthma, and emphysema as well as chronic depression and anxiety. This is excluding other physical factors such as body mass index (BMI), hypertension, or disability. The relationship between stress and the immune system is particularly complex: while short episodes of stress have been correlated with a temporary strengthening of our defenses, chronic stress is responsible for a generalized weakening of the immune response. This is the scientific confirmation of something that we have all experienced at some point in life: after overcoming a period of prolonged psychological pressure and regaining a sense of peace, we develop a minor illness like a cold. This is our body's price for subjecting it to prolonged stress.

In his 1956 book *The Stress of Life*, Canadian Hungarian endocrinologist Hans Selye defined the medical sense of the word *stress* as indicating a general, nonspecific response of the organism to a negative stimulus. The term, borrowed from physics, had been used by Selye in the past, but only after publishing the book did it enter the common lexicon due to the impact of his work on a wide audience of specialists and laypersons. Known as the "father of stress research," Selye allowed scientists to change their perspective on the study of disease. Until the late twentieth century, each pathology was analyzed separately as a unique phenomenon whose symptoms were considered specific. Selye was the first to notice that there were points in common between the symptoms experienced by patients suffering from different diseases and assumed that there must be a thread that connected them. Fatigue, weight loss, a lack of appetite, the desire to lie down, and the reluctance to go to work were all symptoms found in patients with different diagnoses. He used this observation as the foundation of his research, finally identifying stress as the common factor in the origin of different diseases and coming to understand many mechanisms that would later be confirmed by science. "It is not the stress that kills us, but the way we react to it," he wrote in his book. "Every stress leaves an indelible scar and the body pays for having survived a stressful situation by becoming a little older."

The link between stress and premature aging, with the relatively early onset of chronic disease, has been the focus of considerable research, from the Selye era to today, and is widely documented in many areas of epidemiology and psychiatry. In a 2014 study Elissa Epel, at the University of California at San Francisco, in line with Sapolsky's research, highlighted how our high cognitive abilities, such as the ability to learn, remember, and anticipate stressful situations, can create a state of constant vigilance, which is activated even in the absence of a stressful situation. The intensity of this constant alertness varies from person to person, such that some individuals are more vulnerable to stress than others. This subjective difference is at the center of new research studies, aimed at understanding how much our "inner environment" can favor or hinder adverse biological phenomena, such as disease and premature aging.

STRESS AND TELOMERES

So what about telomeres?

As we have already mentioned, telomeric DNA suffers negative consequences from a state of chronic stress, which is one of the factors that can accelerate its shortening. Cortisol and other hormones released into the blood in response to negative stimuli (real or perceived) create oxidative stress and inflammation, the two biochemical conditions that accelerate telomeric attrition. Let's imagine that to our DNA, stress is like a poison, released in small yet constant quantities: it doesn't kill us immediately, but it makes our cellular environment toxic, favoring the eventual onset of pathologies. This phenomenon is particularly accentuated when chronic stress strikes at a young age.

From the late 1960s until 1989, Romanian dictator Nicolae Ceaușescu imposed a severe demographic policy designed to increase the population in his country, which prohibited abortion and contraception and taxed families without children. The result of this legislation was an increase in births, but a corresponding increase in abandoned infants and children, especially those born into large families or with some type of disability. The regime's propaganda described the phenomenon as an act of responsibility for the parents, who entrusted their children to public structures, believing that the state would guarantee the children better care than they themselves could provide. The country's orphanages were filled with abandoned children, who were forced to live in crowded and crumbling facilities in disastrous conditions.

The world knew nothing about this until the early 1990s, when the regime fell, and the first international journalists and observers entered these institutions and found themselves face-to-face with horror: undernourished children, covered in dirt, frequently sick, and deprived of any form of care, sometimes tied to their beds, and generally treated in a violent and inhumane way. It was a shocking discovery that upset Western public opinion, unaware of what was happening in these places of suffering and abuse. In a BBC documentary, Izidor Ruckel, now an adult, who lived in one of these facilities until the age of eleven, says, "We didn't have compassion, we didn't have feelings or emotions. We just existed, to just vegetate. We were just wild animals that needed to be caged up. This is what we were considered pretty much." The news and images of this desperate childhood caused a wave of emotion and solidarity around the world, leading to new channels for the international adoption of children saved from orphanages. Many, like Ruckel, were welcomed into the United States or other Western countries, and despite difficulties were able to achieve a satisfying life, but for many others, their integration was difficult due to the serious psychological consequences of their internment.

The tragedy of the Romanian orphans mobilized psychiatrists, doctors, and scientists from all over the world to study their physical and mental health conditions in an attempt to treat the damage caused by early trauma. These Romanian institutes remained under observation for many years, but the most problematic situations persisted for a long time. Even in 2000, conditions in many orphanages there were not that different from those in the Ceauşescu era. It was that year that pediatricians and scholars from Harvard and other US universities, in collaboration with the Boston Children's Hospital, started the Bucharest Early Intervention Project, a long-term study aimed at investigating the effects that serious and prolonged conditions of deprivation and abandonment have had on these children's health. In particular, the telomeres of 136 orphaned children residing in six different facilities in Bucharest, aged between six months and two and a half years, were observed. Some of these children were entrusted to the care of foster families, while the rest continued to live in orphanages. At the end of the first period of observation, it was seen that the telomeres of the group in the institution were more worn than those of the group in foster care, thereby confirming the initial hypothesis that from the first years of life, conditions of material and emotional deprivation, resulting in prolonged psychological stress, can cause early cellular damage. Life in the orphanage proved to be more damaging than the experience of foster care, underscoring the impact of emotions on our DNA's integrity.

PROTECTING TELOMERES FROM STRESS

What's most striking about telomere studies is that this type of cell damage is irreversible, as mentioned earlier, and even if the consequences can be mitigated by adopting protective strategies, they cannot be neutralized. When the damage is done, it's done; there is no remedy. That's the most important lesson gained from this research: we must protect our health as much as possible because what we lose can never be recovered. Let's commit ourselves right now to improving the quality of our lives, starting with a healthy diet, active lifestyle, and even minimal meditation practice in order to gradually heal our own relationship with the world. Let's treat others with kindness, admit our mistakes, apologize, forgive the wrongs done to us, and relieve the burden of our guilt and pain poisoning our mind and cells. Let's address the mechanisms of stress as soon as possible. Let's not hide behind the excuse that it doesn't depend on us; we can't eliminate stress from our lives, but we can change the way we respond to it. Let's keep repeating this to ourselves. We can choose to ease tensions in the family and at work rather than increase them. We can stop being unnecessarily angry, holding grudges over minor incidents, and storing up resentments. It will only wear us out and leave irreparable damage behind. Our DNA is asking us to do this. What we see in the laboratories is a desperate cry from our cells, asking to be protected from stress, junk food, and a sedentary lifestyle. Let's spare ourselves and others any unnecessary psychosocial pressure because it leaves its mark on our bodies, makes us age more quickly, and makes us sick.

When we talk about longevity we are not referring only to a purely quantitative datum. It is not enough to live long if that extra time we gain is not quality time. Let's make our lives longer and richer, keeping our bodies younger and healthier. This will allow us to better enjoy the years of old age, to enjoy the company of family and friends, our hobbies, and new projects that we haven't had the courage to undertake before. Science doesn't claim to give eternal recipes or definitive answers but instead offers advice based on long and rational observation. Like the simple rule of having a purpose, pursuing a goal that makes us feel good and doesn't harm anyone is one of the great secrets to aging better.

In 2011, the case of Leo Plass, who graduated from the University of Eastern Oregon at the age of ninety-nine, made headlines. The real decline of an individual begins when they lose interest in things and end up repeating their days without a purpose. But it's not about going to the moon or climbing Mount Everest. A little interest like gardening, reading, playing a musical instrument, learning a new language, taking a trip to a new place (even one nearby), creating a puzzle, or volunteering may be enough. Whatever goal we set for ourselves, be it small or large, we are already feeding our brain and body with positive energy, which will keep our mind active and also be good for our physical health. Aging is an inevitable biological process, but living it well can make it more enjoyable.

A study by Harvard University, also reported in the *New York Times*, explained the mechanism by which chronic stress can accelerate the appearance of white hair, even at a young age. Hair pigment is produced by stem cells—that is, progenitors of actual cells—found in the follicles at the base of the hair. From the stem cells come the melanocytes, responsible for hair coloring. Stress stimulates the sympathetic nervous system, which is responsible for the fight-or-flight response and connected directly to the hair follicles, producing norepinephrine, which binds to melanocytes and carries them away from the hair. The result is that the pigment no longer reaches its destination and the hair grays faster. For some years now, scientists studying the link between stress and premature aging have been showing the public two photos of Barack Obama, one from 2008, on his arrival to the White House, and one from 2012, when he was reconfirmed for his second term. The amount of white hair that appeared in just four years is astounding—a sign that the great responsibilities, with the load of stress they entail, can play a nonsecondary role in accelerating our body's aging. A similar mechanism has been found in studies about hair loss due to stress. This is also an experience that many have had in life, and scientific research has confirmed that stress hormones interfere with the normal follicle function, probably contributing to accelerating hair loss in both men and women.

TRAUMATIC EVENTS AND CELLULAR HEALTH

The attrition that stress triggers in our body seems to have the same underlying characteristics in small episodes of daily anxiety as it does in major traumas. In 2018 at Harvard University, in collaboration with the US Department of Veterans Affairs, we completed a study on the cellular consequences of post-traumatic stress disorder (PTSD) in a population of 453 individuals of all genders, primarily war veterans, but also some civilians, all exposed to traumatic situations. PTSD involves intense emotional and psychological suffering, with symptoms such as profound mood instability, nightmares, flashbacks, hypervigilance, exaggerated startle responses, poor reactivity to external stimuli, and a preoccupation with the traumatic event. First included in the DSM-III's official list of diagnoses in 1980, PTSD has historically been associated with the experience of military combat, but was subsequently studied outside this limited field.

This group of people's DNA showed some variability in the state of telomeric attrition. The association between PTSD diagnosis and telomere length was especially correlated with age: the older people in the observation group had shorter telomeres than the younger ones, but to a much larger extent than would occur naturally. Knowing the average rates of telomere shortening in the various stages of life, we can verify whether this is attrition is occurring at natural or accelerated rates. At the cellular level, PTSD seems to have a more marked incidence with advancing age, as if older people have a greater "sensitivity" to traumatic events. By dividing the veterans into categories based on character traits, we also saw that stronger and more resilient personalities are associated with longer telomeres—a sign that individual psychological resources can play a protective role for DNA. This research therefore spotlighted how highly traumatic events can negatively impact our cellular health and promote senescence, but subjective conditions, such as age and personality, can mitigate those effects.

In biology, this subjective ability to resist the negative action of an external stressor is called *resilience*, a complex and fascinating topic that has attracted great scientific interest in recent years. Scholars, from epidemiologists to psychiatrists, have wondered how it works, what characteristics a person must have in order to activate such a response, and whether resilience is innate, acquired, or combination of the two. In 2004, George Bonanno, a professor of clinical psychology at Columbia University in New York City, defined resilience as the ability to maintain a normal state of equilibrium during a highly traumatic event. This definition stems from the observation that in the aftermath of a traumatic event, some people will develop PTSD, while others won't. What is different between these two groups of people? What resources do the resilient people have that the vulnerable are lacking? The answer is complex, and research is still ongoing, but what has been observed so far suggests that there is a genetic predisposition mixed with environmental factors such as life experiences, personal beliefs, and reference values. Each person carries a varying amount of both natural and acquired "antibodies" to stress, and can activate either effective responses to adverse stimuli or insufficient responses that can pave the way for the onset of PTSD.

In order to better understand the susceptibility of various human groups to PTSD in recent years we have begun to broaden the range of research. At Harvard University, after the significant results of the studies on veterans, we wanted to investigate these correlations in an all-women population, selecting subjects exposed to traumatic events in the context of civilian life. Also, in this case we found that women diagnosed with PTSD presented accelerated cellular aging due to telomere wear and tear.

In recent years, thanks to the increasingly detailed knowledge of telomeres and their mechanisms, we have compiled a significant amount of data to confirm the correlation between stressors and premature aging, chronic disease, and reduced life expectancy. We finally have evidence of what happens to our DNA when we are faced with adversity, and now know the biochemical reactions that are triggered and indelible traces they leave in our bodies. Anxiety and phobias play a role like that of stress on DNA, triggering telomere shortening mechanisms. Phobic anxiety is classified as an anxiety condition. A subpopulation of over five thousand women from the NHS was the subject of this specific observation, which detected shorter telomeres in those diagnosed with the condition than in those who were not. Thus oxidative stress and inflammation, causes of accelerated telomere shortening, are then aggravated by the comorbidity of a phobic condition with other conditions, such as depression and substance abuse.

A study published in 2018, conducted by several universities and health institutions in the United States, looked for a correlation between depression and accelerated telomere shortening. A population of 117 individuals aged eighteen to seventy with a depression diagnosis observed for just two years showed significantly shorter telomeres than those seen in people who didn't suffer from depression, hence strengthening the hypothesis that depression is a factor in accelerating biological aging.

RELEASE THE PRESSURE

The skills we have today tell us that being subjected to chronic stress conditions means a high cost in terms of both physical and mental health, quality of life, and longevity. Telomeres give us the exact measurement of the wear and tear that life's adversities exert on our body, at the same time allowing us to take action to limit the damage. A clear message comes from the tips of our chromosomes: excessive psychological and social pressure, especially in childhood and early youth, can be as bad for our health as smoking cigarettes or being obese. Now that we have the biological evidence, it's time to do something to help us get better.

3 OPTIMISM

IMMACULATA DE VIVO

> The optimist sees opportunities in every danger, the pessimist sees danger in
> every opportunity.
> —popularly attributed to Winston Churchill

On September 13, 1940, Luftwaffe planes dropped several bombs on the city of London, and one of them hit the royal residence at Buckingham Palace. The bomb pulverized some rooms in the palace and blasted large holes in its walls. King George VI and Queen Consort Elizabeth later went to the bombing site to inspect the damage. With irresistible British humor, the queen exclaimed, "I am glad we have been bombed. Now we can look the East End in the eye!"

The ability to find a positive side even in the midst of adversity is a typical characteristic of optimists. But what does it mean from a scientific standpoint to be optimistic or pessimistic?

OPTIMISM AND PESSIMISM

Optimism is a psychological attribute characterized by the expectation of positive events. The optimist is confident that tomorrow will be better than today and often convinced that they can control the outcome of certain situations. It's not just reacting well to adversity but about a completely different way of interpreting reality too: what is negative for others is not negative for the optimist, who sees each event from a radically different perspective. Scientific research identifies a strong correlation between optimism and living longer, healthier lives compared to those who are more pessimistic. It seems that having a positive outlook on life and the future helps us to suffer less from chronic disease and avoid premature death. We have evidence for at least three different mechanisms through which optimism affects longevity: increasing the likelihood of engaging in healthy behaviors, slowing down the biological aging processes, and promoting emotion regulation and response to stressful experiences.

Optimism therefore plays an important role in self-regulation because a general confidence in the future allows us to pursue our goals, manage difficulties, and adjust our expectations if certain objectives are no longer achievable. It leads us to create better social relationships because we have a more confident, trusting attitude toward others. As such, it is an effective tool for adapting to life challenges because it alters the way we process and interpret daily stressors, mitigates our emotional responses, and enables us to perceive them in a less threatening way. This "emotional reactivity" is highly individual, but certain factors such as age can help mitigate it: older people have more effective self-regulation mechanisms and tend to have fewer of the emotional changes that are typical of youth. Science confirms the popular saying "age brings wisdom." Greater emotional reactivity is associated with various adverse body conditions such as higher inflammation and a higher mortality risk. On the other hand, strategies that increase emotional regulation are rewarded because changing how we view an event with a strong emotional impact is associated with better cardiovascular health. The optimist tends to see difficulties as challenges rather than as threats and is willing to trade an immediate reward for a long-term benefit. This is why self-regulation mechanisms make optimism an ally of our health: they defuse the potential damage caused by experiencing overwhelming emotional reactions, thus helping us to respond more calmly to adverse events and return more quickly to our baseline.

Obviously, optimism and pessimism are not black and white; between these two extremes, there is an infinite series of intermediate positions. A person who tends to be optimistic can be more optimistic in one circumstance and less so in another, depending on the variables that influence their judgment on each event. For researchers, it's more useful to analyze the extremes because they can attribute to each personality type those attitudes and psychological resources that most characterize it.

Martin Seligman, a famous US psychologist, speaks of "style of attribution" in relation to how people explain events. Let's imagine a relatively minor negative event such as failing a school exam. The first level of interpretation is personalization: Was it my fault or due to an external factor? The second is permanence: Is this an event that will persist over time, or can I do something to prevent it from happening again? The third is pervasiveness: Is this an event that affects all aspects of my life, or are its effects limited to a particular sphere? The pessimist will look at the event as something personal, permanent, and pervasive: it's my fault; it will always happen; I am a failure. The optimist, on the other hand, will interpret it as something external, transitory, and limited: it's not my fault; next time it will be better; this was just an unfortunate case.

This different way of interpreting reality was at the center of a series of neuroscientific tests conducted in 2011 by an Anglo-German team, which tried to understand how optimists manage to maintain a positive vision of the future even in the face of overwhelming evidence to the contrary. The answer seems to lie in the mechanism through which these people welcome and integrate new information into their overall worldview: positive news, which can change their future vision for the better, is widely accepted, while negative news, which could make it worse, is only partially integrated. Particularly optimistic people would therefore tend to select information on the basis of its meliorative or pejorative potential, largely excluding that which is undesirable from their vision of the future.

MEASURING OPTIMISM

In his 1991 book *Learned Optimism*, Seligman offers a test to assess one's level of optimism. The questionnaire consists of forty-eight questions, each with two possible answers. Unlike the many similar tests one can find in newspapers or on the internet, Seligman's is scientifically valid and used to identify the baseline of a subject who is learning optimism. Often the answers differ in some nuances, so the recommendation is to accurately visualize the situation, especially those in which we have never really found ourselves. Here are some examples of questions and answers.

You and your partner make up after a fight:

- I forgive them
- I usually forgive

You get lost while driving to a friend's house:

- I should have turned, but I got it wrong
- My friend gave me inaccurate directions

(continued)

You apply to a local public office and win:

- I have spent a lot of time and energy on the campaign
- I work hard in everything I do

    As early as 1985, Charles Carver and Michael Scheier had developed the life orientation test (LOT), revised and updated in 1994 as LOT-R. This test is still used to assess a subject's levels of optimism, and was used in the Harvard studies on the link between optimism and telomeres. The LOT-R includes ten questions, which are scored using a complex method that crosses the value of the question with that of the answer.

1. Under uncertain circumstances, I tend to expect the best
2. I easily relax
3. If something can go wrong, it certainly will
4. I am always optimistic about my future
5. I enjoy my friends' company very much
6. It's important for me to keep busy
7. I hardly expect things to go as I would like
8. I don't get angry easily
9. I rarely expect good things to happen to me
10. Overall, I expect more good than bad things

OPTIMISM AND HEALTH: FRIENDS FOR LIFE

Being optimistic or pessimistic is not just a psychological trait or interesting topic of conversation; it's biologically relevant. There is an abundance of scientific literature on the correlation between optimism and health that allows us to consider this mental disposition today as a tool for preventing disease and aging. Optimistic-minded people are associated with various good health indicators, particularly cardiovascular, but also pulmonary, metabolic, and immunologic. They have a lower incidence of age-related illnesses and reduced mortality levels. Optimism and pessimism are thus not arbitrary and elusive labels. On the contrary, they are mindsets that we can scientifically measure, placing the attitude of each person observed on a scale that ranges from optimism to pessimism. Framing the baseline of each subject in this way, it was possible to verify the correspondence between optimism level and the relative health conditions.

In 2019, a review published in *JAMA Network Open* by Alan Rozanski, a cardiologist at Mount Sinai St. Luke's hospital in New York City, compared the results of fifteen different studies for a total of 229,391 participants. Those participants with a more optimistic attitude were associated with a lower incidence of heart attack and other cardiovascular diseases as well as a lower mortality rate. Rozanski pointed out that the most optimistic people tend to take better care of themselves, especially by eating healthily, exercising, and not smoking. These behaviors have been found to a much lesser extent in the most pessimistic people, who tend to care less for their own well-being. But the damage produced by pessimism is also biological: the continuous wear and tear caused by elevated stress hormones like cortisol and noradrenaline leads to heightened levels of body inflammation and promotes the onset of disease. Moreover pathological pessimism can lead to depression, considered by the American Heart Association as a risk factor for cardiovascular disease.

The same correlation has been identified in relation to minor illnesses like the common cold. A 2006 study outlined the personality profiles of 193 healthy volunteers who were inoculated with a common respiratory virus. Subjects who expressed a positive attitude were less likely to develop symptoms of the infection than their counterparts with less positive attitudes. Optimism, then, is one of the most interesting nonbiological factors involved in the mechanisms of longevity because it correlates an individual's psychological attributes with their physical health. In this sense, it offers us a further strategy to protect our health.

Optimists tend to live longer, as revealed by research led by Lewina Lee at Harvard University analyzing 69,744 women from the NHS and 1,429 men from the aging study of the US Department of Veterans Affairs. The results tell us that optimists tend to live on average 11–15 percent longer than pessimists and have an excellent chance of achieving "exceptional longevity"—that is, by definition, an age of over eighty-five years. For optimistic women, the likelihood of reaching this milestone is 50 percent, and for men, it's 70 percent. These results are not confounded by other factors such as socioeconomic status, general health, social integration, and lifestyle because, according to Lee, optimists are better at reframing an unfavorable situation and responding to it more effectively. They have a more confident attitude toward life and are committed to overcoming obstacles rather than thinking that they can do nothing to change the things that are wrong.

A survey conducted in France in 1998 hypothesized a correlation between the death rate and group events that inspire optimism. On July 12 of that year, at the Saint-Denis stadium, the French national soccer team won the World Cup against Brazil. Data on deaths from cardiovascular events recorded that day show a singular decline compared to the average recorded between July 7 and July 17, but this effect was limited to the male population, while for women it remained about the same. Although it's not possible to establish a causal link, this curious coincidence suggests that the massive injection of optimism after the team's victory may have played a role in the story.

IN THE OPTIMIST'S MIND

Starting from the premise that a fundamental part of life is the pursuit of goals, it has been seen that encountering obstacles to achieving these goals can lead to different results depending on the individual's level of optimism. If the person has a confident and positive attitude, they will try to overcome the obstacle; if they are doubtful that their efforts will succeed, they will tend to let it go, perhaps experiencing frustration from their remaining attachment to this goal, or may become completely disengaged and fail to achieve their goal. Optimism and pessimism posit this mechanism on a larger scale as a mental attitude toward not only a single goal but also the future in general.

Researchers have studied the relationship between these two attitudes and the results obtained during the course of real-life situations. It has been seen that optimists are more likely to complete university studies, not because they are smarter than others, but because they have more motivation and perseverance. And they are able to better manage the simultaneous pursuit of multiple goals—making friends, playing sports, and doing well at school—by optimizing their efforts: showing a greater commitment to priority goals and less commitment to secondary ones. The optimist seems to invest their self-regulation resources carefully, increasing effort when circumstances are favorable and decreasing effort when they are less favorable, but also by doing more when there is a disadvantage to overcome. In a famous 1990 study by Seligman involving college swimming teams, coaches asked athletes to compete at their best. At the end of the competitions, they were given false results about their performance, increasing their speed by about two seconds, which was low enough to be credible, but still enough for the athletes to be disappointed. After a couple of hours of rest, during which they probably mulled over their bad results in the last race, the swimmers were called to a second race, and the results between the optimists and pessimists were significantly different. The pessimists were on average 1.6 percent slower than during their first performance, while the optimists swam 0.5 percent faster. The interpretation of the experiment was that optimists tend to use failure as a goad to do better, whereas pessimists tend to be discouraged more easily and give up more readily.

Results of DNA studies also seem to confirm the idea that optimism is an effective tool for slowing down cellular aging, of which telomere shortening is a biomarker. This research is still in progress, but the early results are informative. In 2012, Blackburn and Epel at the University of California at San Francisco, in collaboration with other institutions, identified a correlation between pessimism and accelerated telomere shortening in a group of postmenopausal women: a pessimistic attitude may indeed be associated with shorter telomeres. Studies are moving toward larger sample sizes, but it already seems apparent that optimism and pessimism play a significant role in our health as well as in the rate of cellular senescence. Recently, Harvard University scientists, in collaboration with Boston University and the Ospedale Maggiore in Milan, Italy, observed the telomeres of 490 elderly men in the Normative Health Study on US veterans. Subjects with strongly pessimistic attitudes were associated with shorter telomeres—a further encouraging finding in the study of those mechanisms that make optimism and pessimism biologically relevant.

LEARNING OPTIMISM

But if someone is born a pessimist, what can they do? Do they have to resign themselves to a more difficult and perhaps even shorter life just because they have a certain character? Science says no, absolutely not. Optimism is an attitude that can be learned by following paths that can gradually change one's outlook on life. Cognitive behavioral therapies offer promising results because they encourage strengthening coping strategies and containing negative thoughts. We can change our coping strategy by expressing gratitude, practicing acts of kindness, doing meditation, and engaging in visualization exercises. Some research suggests that five minutes a day in which one visualizes the best possible version of oneself is enough to increase one's level of optimism in just two weeks. An exercise in which one must find the most positive possible motivations for various life events has also proved effective, in addition to the numerous cognitive behavioral therapy techniques used to treat depression.

According to Seligman, psychology must not only cure mental suffering and illness but promote the positive aspects of existence too. He is considered the originator of positive psychology, a field that aims to teach people to practice optimism, take care of their well-being, and commit themselves to improving the quality of their lives. Seligman also proposed optimism learning paths for children between ten and twelve years of age based on research findings that an early learning of these strategies allows for a long-lasting protective response against depression and anxiety. Other research devoted to adolescents has correlated optimism with better academic performance.

Optimism is thought to be genetically determined for only 25 percent of the population. For the rest, it's the result of our social relationships or deliberate efforts to learn more positive thinking. In an interview with Jane Brody for the *New York Times*, Rozanski explained that "our way of thinking is habitual, unaware, so the first step is to learn to control ourselves when negative thoughts assail us and commit ourselves to change the way we look at things. We must recognize that our way of thinking is not necessarily the only way of looking at a situation. This thought alone can lower the toxic effect of negativity." For Rozanski, optimism, like a muscle, can be trained to become stronger through positivity and gratitude, in order to replace an irrational negative thought with a positive and more reasonable one.

Indeed, it's important to emphasize that an excess of optimism is as harmful as an excess of pessimism. Positive thinking means that things will improve in the future, not that we are immune from harm, suffering, or bad luck. Our sense of responsibility must remain alert and prevent us from making blatantly risky choices just because we have a positive feeling about the results of that action. Some studies have identified certain situations where optimism was problematic rather than a helpful resource—gambling, for example. Observing the behavior of people who were not compulsive gamblers, it was seen that optimists tended to have excessive confidence in the outcomes of their bets and were reluctant to lower subsequent bets, despite their previous losses.

So there is a range within which our attitudes, optimistic or pessimistic, can be considered healthy. A small dose of pessimism can be lifesaving in dangerous circumstances because it prompts us to seek help when in trouble. It's a fundamental evolutionary mechanism that by making us foresee negative consequences, urges us to be cautious, thus protecting us from avoidable dangers. A harmonious balance between these two extremes, with a strong inclination toward optimism, seems to be the most effective formula.

Being optimistic or pessimistic can influence many subjective experiences, affecting the quality of life. Many scientific studies conducted on the matter suggest that an optimistic attitude can affect the perception of physical pain, with optimists showing more positive reactions to the placebo effect. When taking inactive substances that they believed were painkillers, optimists showed greater relief, whereas pessimists, tending to stress the pain, showed a lower placebo effect. Experiencing a disease is also accompanied by higher levels of anxiety in pessimistic people, who are more likely to develop suicidal thoughts when they feel they have become a burden to others. In a range of adverse situations, optimists tend to be more focused on solving the problem. When losing a job, for instance, they are able to maintain a higher level of personal satisfaction, also linked to a more marked perception of receiving support from their family.

So let's not give up if we're born pessimists. Optimism, strongly correlated with good health and longevity, is a mental disposition that we can learn and cultivate, making it a powerful ally on our path to well-being.

4 FORGIVENESS AND GRATITUDE

DANIEL LUMERA

Wear gratitude like a cloak and it will feed every corner of your life.
—Rumi, *Rumi: The Path of Love*

A NEW IDEA AND EXPERIENCE OF FORGIVENESS

Why is forgiveness such an important element in the quality of our life and relationships as well as the process of social transformation? What impact does it have on psychophysical health?

The idea of forgiveness that will be presented in these pages completely redefines the experience of it. Aside from the common meaning of the word, which would have no reason for existing without guilt, error, sin, abuse, snub, or offense, it's possible to explore forgiveness through a new paradigm, which goes beyond the victim-tormenter duality and enters an experiential field different from common logic. After more than fifteen years of research in this area, the International School of Forgiveness was founded in 2013. It is a research and training laboratory that offers its methods and protocols to schools, prisons, hospitals, and peace processes, entering the macro areas of education, justice, and health.

BEYOND LONELINESS

More and more people are experiencing a profound feeling of loneliness, isolation, and separation from the world. Paradoxically, many of these people feel alone even when they are in a crowd. It's an intimate feeling, exacerbated by a context that promises to end social isolation, but has achieved the opposite result. This deep feeling of loneliness and separation is certainly influenced by many factors (biological, relational, social, urban, technological, emotional, cognitive, and behavioral), but if we have the courage to really listen to ourselves, to descend into the abyss we have inside, then we could discover that behind it there's an ancient nostalgia: ourselves.

We feel alone because we are lost: we have lost the sense of ourselves and therefore the sense of who we are in relation to others. That deep contact has failed, that state of integrity, wonder, clarity, and enthusiasm, to live a flourishing life. The word *enthusiasm* comes from the Greek *en*, "inside," and *theós*, "god"—that is, having a god (the infinite, the universe) inside.

It's the awakening of that unstoppable force we experience when we feel connected and united to life: the powerful awareness that "everything" is within us. It is being one with every living being, in joy and pain, birth and death, and above all love, and maintaining a deep sense of serenity and peace even through the most stormy experiences of life.

Loneliness hides the seed of self-nostalgia, and thus forgiveness, if correctly understood and integrated, can transform this lack into fullness and gratitude. Day after day, without even realizing it, we build a wall around our heart—a wall made of anger, guilt, shame, judgments, recriminations, and rigidity. In this way, we have moved away from ourselves and others, imprisoning ourselves in the weight of the past and anxiety of the future. Forgiveness is the purifying water that drop by drop, "digs into the rock," dissolves every wall and washes away pain, freeing the heart, giving new light to the eyes, and transforming every present moment into a gift.

AN INNER PLACE

Perdonare ("forgiving" in Italian) comes etymologically from *per*, "completely," and *donare*, which means "giving," the giving par excellence, or transforming everything that happens to us (inside and outside) into a gift: an inner place where we can experience the power we have over what we feel, perceive, and think; an intimate space where one can feel firsthand that a liberation from the instinctual impulses of revenge, resentment, and guilt is possible. When a change occurs in our inner world, the external one also changes as a result. We are like radios that transmit on certain frequencies and communicate through them. When we are tuned to the station of anger, then we play along with people, situations, and circumstances that broadcast on the same frequency. The instrument of forgiveness is used to tune one's "cognitive, perceptive, and emotional frequencies" to channels that lead to experiences of happiness, prosperity, and harmony.

A ZEN STORY

A pupil went to their teacher and presented them with a problem:

"Master, I live moments of irrepressible anger. How can I heal?"

"Show me what it is," said the master.

"Well, off the top of my head, I can't show it to you," replied the neophyte.

"When will you be able to show me?"

"It comes out when I least expect it," the student replied.

"So," concluded the master, "it's not your true nature. If it was, you could show it to me at any time. When you were born, you didn't have it, and your parents didn't give it to you. Meditate on it."

RADICAL FORGIVENESS

I ask your forgiveness.

Forgiving those who have hurt us and the pains of the past seems like an extreme challenge only fit for a few people. What if we dared much more? What if we tried to look our worst enemies in the face—a murderer, pedophile, or serial killer serving a life sentence—and eye to eye, we had the courage to utter four simple words: "I ask your forgiveness"?

Four simple words: "I ask your forgiveness."

Even just thinking about it would probably trigger a deep sense of rejection and anger in most people. Yet that's exactly what I do with the inmates I meet in prisons. I ask forgiveness from each of them. Truthfully. The tears that have fallen from their eyes and mine tell without pretense how far we still are, as a society, from the possibility of understanding the roots of others' discomfort and pain. A total lack of awareness emerges every time someone says, "Let's put them in a cell and throw away the key." And in doing so, they condemn themselves and the people they love to ignorance.

Every time I looked a prisoner in their eyes and said, "I ask your forgiveness," I found reflections of my own judgments on their faces, including my condemnations, rejection, anger, helplessness, misunderstanding, and hate. In those faces, I saw the failure of a society that can't and doesn't want to understand the root of pain and discomfort. We prefer to condemn and punish what we fail to understand rather than take full responsibility for our ineptitude as human beings.

Inside prisons, I met many people who were there because they had not had a real chance in life: they had not received attention, love, parenting, care, or affection. Everyone deals with their pain the best they can, based on the tools they have received from life. I ask for forgiveness for the lack of empathy; I ask for forgiveness for all the times we fail to understand situations, to detect a request for help and love behind anger and pain; I ask for forgiveness for the insensitivity.

What happens when a dog is injured and we come closer to help? They bite, but not because they're bad. They bite because they feel pain and are trying to protect themselves. Punishing and beating the dog is the solution for those who don't understand the origin of their pain.

I asked for forgiveness for the absence of love, compassion, awareness, and understanding; I asked for forgiveness for not being able to see beyond anger. Beyond the pain. Beyond the wounds. I asked for forgiveness for not being able to cure that discomfort at its source. For having abandoned, rejected, or hated. I asked for forgiveness for not having understood that behind every behavior, even the most violent, there is always a request for love and help.

"We can be excused for not finding a solution for all of this."

Many of us criticize or condemn what we fail to understand. That's also why I asked for forgiveness. Because each of their faces was mine as well. We are all part of one whole. Our lives are deeply interconnected, and if someone ends up in jail, the blame is not solely on that one individual; we all bear some responsibility. If an educational system fails, then society as a whole will also fail. Can we—or are we willing to—develop a new sense of responsibility? Where each of us feels connected with and dependent on one another and the world? I don't know. But I do know that in order to understand, I must begin with these four simple words: "I ask your forgiveness."

FORGIVE THE UNFORGIVABLE

The life of Eva Mozes Kor is a clear demonstration of how forgiveness, when authentic, contains an extraordinarily liberating and transforming power. In 1944, as a child, she was deported from Romania to a concentration camp in Poland, Auschwitz, where she managed to survive Josef Mengele's experiments on twins. As an adult, then an essayist and naturalized US citizen, Eva decided to forgive all the Nazis, even those who indirectly participated in extermination. This was a strong, subjective, and personal choice. She wrote in 2017 that the time had come to move on; that it was time for the wounds to heal. The time to forgive, but not forget.

How is it possible that a person who has been tortured, humiliated, and deprived of her dignity and freedom, and looked into the eyes of mass murderers, can choose to forgive? How could this happen? What took place inside her to lead her to such a radical decision? She describes it like it was an enlightenment. Suddenly she discovered that she had a power: that of forgiving. The power to decide about her life. Forgiveness was her power, and she could use it as she wanted. It was a surprising discovery. Choosing forgiveness allowed her to regain her freedom and the joy of living because a victim has the right to be free, but they cannot be if they don't shake off the weight of pain and anger. Eva had realized that the role of the victim had made her lose control over her life, tied as she was to her jailers through a cord made of hatred, hurt, helplessness, and anger that she kept feeding with painful memories. She understood that those who wear those clothes always and only act in response to what others say or do.

Forgiveness allows you to reclaim your life. "I have this strength." It's not a matter of not reacting but instead acting free from hatred, resentment, and powerlessness.

Every time, for one reason or another, the ghosts of the past reemerged, Eva felt herself collapsing. The hatred was still there. Intact, exhausting, with all of its destructive force, because the greatest victims of hatred are those who harbor it within themselves. With the same force with which anger had imprisoned her in the past, she felt that forgiveness pushed her to rediscover life. Eva became passionate about her present and all the possibilities she had to do something useful in the future. She realized that by giving up the grudge, she had found a new way to relate to her wounds.

Rather than letting them dictate the rules, she decided to accept this power and chose to feel peace as opposed to hate. Discovering forgiveness has given her back the greatest gift that her mother had passed on to her: the ability to be happy, despite everything.

So, it is possible to free oneself, but forgiving doesn't mean forgetting. It's not about removing from one's memory everything that happened. On the contrary, it means keeping in mind what happened by understanding its teachings.

Our instincts suggest that hatred, punishment, anger, and the desire for revenge are natural protection mechanisms, while forgiveness, on the contrary, makes us vulnerable because it's perceived as an act of weakness that denies our experiences. But forgiving is not, in fact, justifying or condoning. Nor is it a superficial denial strategy motivated by the desire to resolve a conflict; it's not about false courtesy and good manners that hide veiled accusations. It is certainly not an act of humiliation, denigration of oneself, or loss of dignity in the hope of pitying and provoking a sense of guilt.

Many might think that with these acts of remission, Eva had denied the atrocities of the Holocaust. But how could she? Her concept of forgiveness does not remove memories; it eliminates their burden. Forgiveness is for us, to free us from the chains that bind us to them. In the last twenty years of her life, since she learned to forgive, Eva said that she could breathe better, but the most important thing is that she could examine the past, remember it in every detail, and talk about it without being overwhelmed. This gave her the the energy to carry out much-needed initiatives to keep memory alive, not to be exploited to foment resentment toward those responsible for the Holocaust, but on the contrary, to prevent hatred from fueling other abuses, wars, and genocides. And those who have been hurt by the atrocities that still shake the world today should be taught forgiveness. As a weapon of self-healing and seed of peace. For ourselves and our society.

Whoever encounters forgiveness, the authentic kind, in their own life, whoever chooses it with courage, inevitably transforms themselves and is liberated, rediscovering the kindness and lightness of heart that have been long lost. It seems impossible, but it happens.

The forgiveness revolution is an evolutionary strategy that allows you to live a better, healthier, and happier life, free from the pain of the past and uncertainties of the future. But above all, it makes you the architect of your present—the only time in which we can choose life and love.

SEVEN LIFE-CHANGING STATEMENTS

We can summarize forgiveness in seven statements, or rather, in the intention and meaning of these seven affirmations.

FIRST STATEMENT: I FORGIVE ME, I FORGIVE YOU

I forgive myself and I forgive you because together we allowed this to happen. In this first statement, I become aware that what happened depends on the existence of both you and me, and that without either one of us, it would never have happened; I abandon a dual concept of forgiveness, which is based on the victim-tormenter polarity, and implies that one of the two made a mistake and therefore there is a fault to atone for by only one person. This simple statement entails a profound assumption of responsibility by both people involved, and allows us to go beyond the dichotomy of victim and tormenter, right and wrong. By both parties accepting responsibility, the details of the offense no longer matter, being right or wrong no longer matters.

As already stated, if we imagine ourselves functioning as radios that transmit on certain frequencies, we will likely meet those who are also tuned into the same wavelengths. If we think in these terms, it will not be difficult to take responsibility for things that happen in our life. If we decide to embrace this perspective, then we will be facing a crossroads: feeling guilty for what happened, as authors (albeit unaware), or being happy for the good news. Which news? That if we attracted what we're living, we also have the power to change it and stop being the cause of that suffering. Assuming this level of responsibility will give us the key to change: if I accept that I have attracted that situation (or that person), it's true that at any time, I can change frequency to tune in and broadcast on more harmonious frequencies. In this first step, we focus on the assumption of responsibility and root of the causes, and no longer on what specifically happened.

SECOND STATEMENT: I FREE MYSELF, I FREE YOU

Regardless of what happened, all I want is to free you and me from all the causes that create suffering. Can you feel the power of this statement? I free myself, I free you. The only thing I care about is absolute freedom from the causes of suffering. Hatred, resentment, fear, ignorance, guilt, and helplessness. The deep will to free oneself from all the causes that create suffering translates into determination. Even in the second statement, it no longer matters who is to blame, who is the victim and who is the tormenter. We become fully aware that in this very moment, we have the possibility to free ourselves from whatever is creating suffering: thoughts, memories, feelings, sensations, images, and emotions. If uttered with sincere intention, this statement has the power to make us focus on our real goal: to free ourselves from all the causes that create suffering. It is a profound choice that we must make: Do we want to keep trying to be right, or would we rather free ourselves from everything that causes pain to us and others? The liberation from hatred, resentment, powerlessness, and anger can begin right here, right now, if we want it to. A liberation that starts at the root of one's feelings. It's about tuning in to a firm intention of freedom. What do we value more, asserting our reasons or being free?

In this second step, we become aware of the principle of reciprocity according to which it's only by freeing others that I can free myself. The principle of reciprocity allows us to interconnect "freeing you I free me and freeing me I free you." This firm and authentic intention to liberate ourselves can give new and unexpected meaning to what happened in our life.

FREEDOM

An ancient Zen story tells of two monks, Tanzan and Ekido, who walked along a muddy road. Near a village, they met a girl with a wonderful golden kimono, who was trying to cross the river. The girl couldn't move, fearing that her kimono, getting wet, would be irreparably damaged. Without hesitating, Tanzan offered his help, took her onto his shoulders, and carried her across the river. After this encounter, the two monks continued their journey. Arriving at the monastery, Ekido, who had been restless for the rest of the trip, suddenly blurted out. Unable to hold back his anger and in a bitter tone, he turned to Tanzan with these harsh words: "Why did you carry that girl on your shoulders? You know well that our vows forbid us to touch women!" Tanzan, unphased, looked at his travel companion with a smile and replied, "I left the girl a few hours ago, but you're still carrying her on your shoulders."

THIRD STATEMENT: I LOVE ME, I LOVE YOU

Imagine saying these words to a person who has hurt you. What is the fundamental meaning of this third statement? What do we want to achieve? "I love myself and I love you" despite everything that has happened because

I recognize that we are the same, voice of the same voice and light of the same light. It's this common bond that I accept and love. "I love myself, I love you" means that beyond what happened and the roles we play, I am capable of loving me and loving you. These first three statements—"I forgive myself and I forgive you," "I free myself and I free you," and "I love myself and I love you"—contain and express the polarity "you and me." Forgiveness, freedom, and love represent the bridge that restores the original bond that was lost. Whether the object of our forgiveness is a person, relationship, illness, place, or situation, we can use these first three statements to create a bridge between the subject who wants to forgive and the object of forgiveness. It is a bridge built with the bricks of forgiveness, freedom, and love.

FOURTH STATEMENT: THANK YOU

The fourth statement is different from the first three, going beyond the me-you polarity. It has to do with a question that often arises spontaneously: "Thank you for what?" There are endless reasons to be thankful.

Thank you for how we loved each other, thank you for the time we were given, thank you for the love we lived, thank you for every moment that we spent together and will never come back, but also thank you for the pain, thank you for the betrayal, thank you for the abandonment, thank you for all that was. Thanks for everything. The fourth affirmation opens the heart's gates. We will be able to find the courage to say, "I don't know why it happened, but I trust life. Often, the end of a relationship is felt as a tragedy in the moment, but then, with time, it can turn into the greatest fortune. Thanks to that end, how many things did we learn and how many important people have we had the opportunity to meet? It's an unconditional gratitude that is rooted in a fundamental trust in life.

GOOD LUCK OR BAD LUCK? THE ZEN PEASANT PARABLE

Once upon a time, in a Chinese village, there was an old farmer who lived with his son and a cow, their only source of livelihood. One day he discovered that the cow had gotten out of its enclosure and disappeared. While looking for the cow, he ran into his neighbor, who asked him where he was going. When he replied that he had lost his cow, the neighbor responded with a shake of his head, "Bad luck."

(continued)

"Good luck, bad luck, who knows?" answered the farmer and continued on his way.

Beyond the farmlands, next to the hills, he found his cow grazing peacefully beside a magnificent horse. He led the cow back home, and the horse followed him. The next morning, the neighbor came to ask about the cow. Seeing the cow again in the enclosure next to the magnificent horse, he asked the farmer what had happened. When he explained that the horse had followed him, the neighbor exclaimed, "How lucky!"

"Good luck, bad luck, who knows?" replied the farmer and went back to his business.

The next day, his son was discharged from the army. He tried to get on the horse, but he fell and broke his leg. The neighbor, who was passing by on his way to the market, saw him sitting on the porch with his leg in plaster while his father was hoeing the garden and asked what had happened.

He listened and with a shake of his head commented, "Bad luck!"

"Good luck, bad luck, who knows?" answered the farmer, continuing to hoe. The next day the young man's ward marched along the path. During the night, war had broken out and the men were on their way to the front. Seeing that his son was unable to go with them, the neighbor leaned over the fence and remarked to the farmer that at least he had been spared the misfortune of losing his son in the war. "How lucky!" he exclaimed. "Good luck, bad luck, who knows?" the farmer replied and kept plowing the field. That evening, the farmer and his son sat down to dinner, but after just a few mouthfuls his son choked on a chicken bone. At his funeral, the neighbor put his hand on the farmer's shoulder and said sadly, "Bad luck!"

"Good luck, bad luck, who knows?" replied the farmer, placing a bouquet of flowers next to the coffin. A few days later the neighbor came to him with the news that his son's entire battalion had been massacred. "At least you were close to your son when he died. How lucky!"

"Good luck, bad luck, who knows?" answered the farmer and went to the market.

THE TEST

Sincere gratitude has the power to take you beyond the apparent "positive and negative" polarity.

- Think about a significant person
- Remember a good time together
- Place a hand on your heart and give thanks
- Remember a moment of suffering
- Place a hand on your heart and give thanks
- Repeat the exercise several times until you feel gratitude in both cases

---

SHORT PRAISE OF GRATITUDE

The Dalai Lama once stated, "Every day, think as you wake up, today I am fortunate to be alive, I have a precious human life, I'm not going to waste it."

Grateful to that gratitude that has marked all the crossroads of our life. Grateful for everything, from the most intense pain to the greatest love. Grateful for all that happened. Grateful for every breath, every step, every smile, every caress. Grateful for what we were able to eat. Grateful for all the steps we were able to take, the houses we were able to live in, and the water we were able to drink. Grateful for all the things that have passed. Grateful for having had them—every experience. Grateful for the gift of this moment, a treasure of beauty that's fragile and infinite. Grateful for the pain too, for the loss, the difficulties we lived and continue to live. When a person is grateful for what happened in their life, in love and pain, it means that they're fully satisfied with their existence, and that everything that happened has been integrated, understood, and transformed into a gift. Being grateful for a pain means that the pain has been turned into a precious opportunity, its meaning has been understood, and it's been used as a resource to develop perseverance, humility, patience, tenacity, passion, strength, determination, and humility.

FIFTH STATEMENT: ONE IN THE ONE

NASA astronaut Russell "Rusty" Schweickart said,

> You identify with Houston and then you identify with Los Angeles and Phoenix and New Orleans. And the next thing you recognize in yourself is that you're identifying with North Africa. You look forward to it, you anticipate it, and there it is. And that whole process of what it is you identify with begins to shift. When you go around the Earth in an hour and a half, you begin to recognize that your identity is with the whole thing. And that makes a change. . . .
>
> You look down and see the surface of that globe you've lived on all this time, and you know all those people down there and they are like you, they are you— and somehow you represent them. You are up there as the sensing element . . . . And somehow you recognize that you're a piece of this total life.

Schweickart's "recognition" of his unity with the earth as a whole was not only a visceral reaction but a true expansion of consciousness too.

The fifth statement indicates the principle of the fundamental unity of life: it literally means that I'm one with the universe (the one).

It celebrates our condition of original integrity, healing the perceptual fracture that makes us feel separated from the world, things, and others. This statement expresses the indissoluble bond with the whole creation, and celebrates the intimate interconnection, interdependence, and belonging to this fundamental unity. The feeling of oneself beyond all division and separation; being one with the fundamental unity of existence. Where that "self" exists infinitely beyond the sense of one's ego. In the thousand-year-old texts of the *Upanisads* (the set of Indian philosophical texts whose most important parts were probably written between 700 and 300 BC; *upa-nisad* literally means "to sit close," indicating the moment of teaching that took place when the student sat alongside the Vedic master), we find several philosophical statements similar to the fifth affirmation. These statements are quoted under the Sanskrit name of Mahavakya (*maha*, "great," and *vakya*, "said") and are considered the highest philosophical truths contained in the Vedas. The most famous among these are, for example,

- *Prajnanam brahma*: the Absolute is awareness (*Aitareya Upanisad 3.3, Rig Veda*)
- *Ayam atma brahma*: the Self (Atman, the essential identity of each of us) coincides with the Absolute (*Mandukya Upanisad 1.2, Atharva Veda*)
- *Tat tvam asi*: You are That (referred to the Absolute) (*Chandogya Upanisad 6.8.7, Sama Veda*)
- *Ekam evadvitiyam brahma*: the Absolute is one, without a second (*Chandogya Upanisad, Sama Veda*)

These great philosophical revelations were so intensely studied as an object of meditation because they allowed one to reach the highest state of consciousness, beyond self-interest and craving.

SIXTH STATEMENT: ONE IN PEACE

Peace begins with you, from what you are feeling, thinking, and doing right now. It's not a goal to reach but instead the choice one makes to live in this moment. It's a state of inner consciousness from which actions, words, emotions, choices, and decisions emerge. Making war and using violence to achieve peace is like screaming to get silence. It will never be a real solution. The Indo-European etymological root *pak* indicates "to unite, to weld, to bind." It restores the lost bond between man and love. To be "one in peace" means to be one with it, with no more separations created by the mind, no space and time that divide and separate us from experiencing it. Peace is not something to be achieved but rather the natural state of our being when the mind is still and silent. It is not a goal but rather a way of walking and being, the choice I make at this moment.

THE TRAIN STATION

My mind could enter a state of peace, silence, and harmony on command. That's why I believed I had reached maturity in meditation. I talked about it with my master, who after a long reflective pause, gave me a difficult task: to practice meditation for an hour a day in a train station. "The noise you will find outside doesn't matter. Your senses will need to be perfectly controlled and reabsorbed into inner awareness. Places and people bring to the surface what we have inside. If inside you there's only silence and peace, then you will find silence and peace anywhere."

SEVENTH STATEMENT: ONE IN THE LIGHT

Light is intended here not only as a symbol of life and element at the base of creation but also a bearer of clarity and purity. "One in the light" means being one with the light of existence. In the *Isha Upanisad* (verse 16), there is an aphorism that contains another great philosophical statement of the Vedas:

> tejo yat te r) paj kalyanatamaj tat te payami yo 'sav [asau
> purusah] so'ham asmi

Literally translated, it means that "I have seen the light, which is the most beautiful form. I am what you are. I am that." Among all possible interpretations, two are particularly interesting. The first emphasizes that in our eyes, light is formless. Therefore, declaring that it's the most beautiful form is to say that the highest form of beauty among all forms is expressed through something that doesn't have form. The second interpretation puts the stress on the manifestation of divinity. Light is the most beautiful form through which the divine aspect of existence manifests itself to our eyes. One not only can have the privilege of accessing this vision but also realize that we are one with it. *So'ham*, I am that same light. What's interesting about these states of consciousness are precisely the side effects: the condition of peace, well-being, balance, love, compassion, optimism, gratitude, and happiness that we carry as a gift, after having fully realized the authentic meaning contained in these statements, which purify the mind, make virtue flourish, and open the heart. It's a state of enthusiasm and constant wonder in being immersed in the miracle of life, which allows us to see everything with the same intensity as those who experience it for the first or last time, through an enlightened mind.

ONE FORMULA WITH SEVEN STATEMENTS

1. I forgive myself, I forgive you
2. I free myself, I free you
3. I love myself, I love you
4. Thank you
5. One in the one
6. One in peace
7. One in the light

The first three statements aim to build a bridge uniting the sense of separation implied by "me and you"—a bridge built through forgiveness, freedom, and love. The fourth allows us to realize that unconditional gratitude has the power to take us beyond duality and the illusory perceptual fracture that separates us all. The last three statements celebrate and recognize the fundamental unity with the universe, peace, and life.

HOW TO MEDITATE ON IT?

A good method is to repeat these seven statements as if they were a mantra so as to feel and explore their meaning along with their effect on the mind, emotions, and body. How many times should this forgiveness formula be repeated? Until the need is felt. An interesting and symbolic number could be seventy times given that in the Gospel of Matthew (18:21–22), when Peter approaches Christ and asks him, "Lord, if my brother commits sins against me, how many times will I have to forgive him? Up to seven times?" Jesus replies, "I don't tell you up to seven times, but up to seventy times seven." Before starting, you can choose a person or situation that you want to "purify" or better align with the five evolutionary forces contained in the formula: forgiveness, freedom, love, gratitude, and unity. Then you need to sit in a comfortable position and start repeating it. It can be whispered or mentally pronounced.

The formula can also be repeated at any time of the day while walking, driving, or showering.

During the repetition, whatever comes to mind or whatever you feel will simply have to be forgiven, liberated, loved, thanked, and brought back to the unity of being. It doesn't matter what emerges during the repetition, be it pain, anger, or love and joy. What will make the difference will be how we use what has emerged. The task is simple: educate your mind to bring everything that comes to it back to the intention of forgiveness, freedom, love, gratitude, and unity. These five intentions (or forces) must become the only desires and wills living in us. We are so used to judging, rejecting, and condemning what we feel that it won't be difficult to experience a sense of liberation, peace, and awareness with a little training.

How to count the number of repetitions without getting distracted?

It is sufficient to craft or get a necklace made up of seventy balls or pearls, and then slide your finger on each of them at each repetition of the seven affirmations. The mind will have to learn to stay focused on the five intentions whatever arises—pain, fatigue, confusion, doubt, anger, helplessness, guilt or even lightness, clarity, joy, happiness, or love. It doesn't matter what we feel but rather what we do with it: forgiving, freeing, loving, thanking, and celebrating the unity of life.

SUFI STORY

"What is forgiveness?" the student asked the teacher. The teacher smiled, took a stone, and placed it in front of the student: "The violent would use it as a weapon to do harm. The builder would make a brick to build a cathedral on. For the weary traveler, it would be a chair on which to sit and rest. The artist would sculpt their muse's face on it. Anyone who's distracted would trip over it. The child would make a game of it. In all cases, the difference is not the stone but instead the person. Through forgiveness, a person chooses to transform the stones of life into love."

5   THE SCIENCE OF FORGIVENESS AND GRATITUDE

IMMACULATA DE VIVO

> Forgiveness does not change the past but it does enlarge the future.
> —Paul Lewis Boese, quoted in *The Weekly Digest* 53, no. 8

FORGIVENESS, BETWEEN SCIENCE AND MORALITY

Forgiveness is a matter with high moral implications, which calls into question individual sensitivity, the subjective consideration of good and evil, and religious convictions. Starting from the observation that forgiveness is a way to heal emotional wounds, scientists have been interested in trying to understand the consequences that this can have on physical and psychological health.

There is already substantial scientific literature on the effects that practicing forgiveness has on well-being and longevity, thanks to various research institutes around the world. A cross-analysis of the data collected so far shows that forgiveness is associated with lower levels of depression, anxiety, hostility, and addiction to nicotine and other substances, and a higher level of positive emotions and satisfaction with one's life. People who have adopted such an attitude enjoy more social support and report fewer symptoms of physical discomfort. The mechanisms that forgiveness brings into play can be explained first of all by the observed subjects' ability to better regulate their emotions, replacing the negative response—mulling over the hurt suffered or removing it—with an alternative response that lowers stress and defuses the detrimental spiral.

In psychotherapy, various methods have been developed to help people suffering from a wrong to adopt the path of forgiveness as a means of healing. In particular, two models have proved effective: Robert Enright's model and that of Everett Worthington, called REACH.

ENRIGHT MODEL

The Enright model is a treatment involving twenty steps divided into four fundamental phases:

1. Awareness phase: revealing the negative feelings that the offense has caused

2. Decision-making phase: deciding to adopt forgiveness while being aware of the meaning of this action

3. Operational phase: working on forgiveness as a way to understand the person who has offended and establish a channel of empathy with them

4. Deepening phase: understanding the meaning of suffering, feeling connected with others, overcoming the negative effects of the wrong suffered, and formulating new life resolutions; forgiveness can free the person from the "emotional prison" of resentment and anger

WORTHINGTON'S REACH MODEL

Worthington's model is structured in five phases that together constitute the acronym REACH.

- *Recall*: recalling the wrong suffered, the pain, and the connected emotions

- *Empathize*: establish empathy with those who have offended and stay in their perspective to understand the reasons for the action, without condoning the action itself or disregarding one's feelings

- *Altruistic*: carrying out an altruistic gesture by recognizing one's own shortcomings and admitting that others have asked forgiveness

- *Commit*: commit to publicly forgive

- *Hold*: maintain the commitment over time, without uncertainty as well as without returning to anger and bitterness

---

It is generally believed that a therapist's guidance is essential for the full success of such a delicate path, but studies have been conducted on self-help manuals too, in which the person proceeds independently. In particular, a 2014 US study noted that a six-hour course of solitary therapy through a manual based on the REACH method actually increased the individuals' forgiveness ability, leading scholars to consider these books as valid tools to support and integrate traditional therapies.

A 2014 review of fifty-four different studies on the effectiveness of forgiveness therapies showed that participants following these psychotherapy paths had a significant increase in their ability to forgive compared to those who had not engaged in the same treatment or had received other types of support. The effect of these therapies on health was linked to a significantly lower level of depression and anxiety along with an increased sense of hope. These results encourage the adoption of forgiveness as a tool to treat the ailments caused by an offense.

How effective forgiveness is was the core of a series of studies whose results were collected in a 2018 article by University of California at San Francisco researchers who investigated the role of forgiveness in healing "interior wounds"—that is, various types of psychological trauma reported by soldiers returning from war. These people went through extremely hard experiences and then have to deal with their sense of morality, tested by the actions they have been forced to perform. From the reports of the therapists who have followed these delicate cases, researchers have identified the common traits of challenges related to inner wounds. On returning from war, veterans often feel ashamed, alienated, and disillusioned, and in some cases they question their own worth as human beings. Those who have fought on the field feel that war has awakened within them a "dark side," which they define as "the beast" or "the monster." It's an entity that feels alien and that they carry with them even after returning home. It prevents them from still perceiving themselves as a good person, caring parent or spouse, and loving and kind friend. These types of injuries often lead veterans to self-harming attitudes that can last for years or even decades without them fully understanding their condition. Some go as far as sabotaging their relationships, jobs, or any other possible source of happiness, feeling that they deserve nothing nice and satisfying in life. In other cases, they feel emotionally numb, unable to experience balanced emotions, or on the contrary, they fall prey to seemingly unmotivated outbursts of anger and despair. In the most serious cases, they decide to isolate themselves from any type of intimate relationship, and avoid contact with the people and things they once loved, frequently getting lost in drug or alcohol addiction. Some consider suicide, while others eventually follow through on it.

Using forgiveness as a healing path has supplied important results, but what does it mean in this context, where the person to be treated is the one who has inflicted the wrong and not the one who has suffered it?

Researchers clarify that it's important for veterans to forgive themselves first, reconcile with their morality upset and wounded by what happened, and regain possession of a compromised mental balance. Sometimes one seeks the forgiveness of the people who have been wounded or killed in combat, at other times divine forgiveness or that of a higher entity, while at other times the forgiveness of loved ones who they've been estranged from after returning home. Whatever the cause of remorse, forgiveness toward oneself is key to the most effective therapeutic pathways. According to the studies, this path may not be sufficient in itself to heal inner wounds, but it has proved useful in facilitating this process, helping people find a serenity on which to build a more lasting healing along with a healthier relationship with themselves and others.

In a 2015 study carried out in collaboration with several US universities, the relationship between self-compassion, which can be considered a form of forgiveness toward oneself, and PTSD symptoms was investigated in a population of 176 women aged between eighteen and sixty-five. Starting from the 2013 World Health Organization's data estimating that 28 percent of women globally experience at least one episode of interpersonal violence in their life, researchers analyzed the attitude of these women diagnosed with PTSD and their mental health, noting that greater self-compassion is linked to a lower likelihood of developing PTSD symptoms along with a greater ability to manage stress and negative emotions.

Science too must often ask itself moral questions when it comes to studying phenomena that involve ethically strong decisions. In a *American Journal of Public Health* article in February 2018, Harvard University epidemiologist Tyler J. VanderWeele questioned whether forgiveness is always morally acceptable. In order to answer this question, the scientist underlined how forgiveness must be distinguished from other similar but different actions, such as condoning, reconciling, forgetting, tolerating, justifying, not demanding justice, and excusing. Forgiveness starts from the idea of wanting the good of someone else without apologizing or forgetting the bad action suffered and continuing to desire a settlement of the wrong. As an example, VanderWeele reports the case of a twenty-year-old man who got drunk and vandalized a mosque in Fort Smith, Arkansas, in 2016. The boy asked for forgiveness, and the mosque's community of believers granted it to him. This didn't mean that his actions had been justified or excused but only that the believers didn't want to ruin his life in any way and asked for a lenient sentence. Keeping these distinctions in mind, the case is an illustration of what it means to forgive but also what it does not mean: the victims didn't deny the harm that had been done to them, its implications, and the feelings that derived from it, yet nevertheless choose forgiveness—that is, replacing a bad feeling toward the guilty with a positive feeling of compassion.

According to VanderWeele, given these conditions, forgiveness is always morally acceptable, and is granted even if the offender hasn't repented and doesn't ask to be forgiven because it's something other than reconciliation. It's something right in itself and an act of love toward those who have hurt us; it's not necessarily an exchange. The evil done and suffered is not denied, thus preserving the victim's respect for themselves as well as the culprit as a human being who made a mistake, to whom a chance for reflection and change is conceded. Ignoring the option of forgiveness as a therapeutic path means abandoning people to resentment, leaving them trapped in the cage of a grudge that can undermine their health. Scientific evidence now tells us that forgiveness promotes health, mind and body well-being, and the salubrity of human relationships. It's a tool for the victim to free themselves from the past as well as a resource to break that form of "dependence" that still binds them to the wrong they suffered and the person who committed it, thereby fostering an attitude of compassion, acceptance, and harmony in human relationships.

GRATITUDE IS GOOD FOR YOU

Cultivating gratitude and practicing it costs nothing, and has tremendous benefits on multiple levels. Scientific research on this matter has quadrupled in the last ten years, and recognizes gratitude as one of the fundamental keys not only for emotional and psychophysical well-being but also for creating more satisfying relationships, working better, and enjoying better health. Let's see the main benefits of a simple "thank you."

1. It opens the door to more relationships. According to a 2014 study published in *Emotion*, saying "thank you" is not only a sign of good manners; showing appreciation can help you find new friends. That study found that thanking someone newly met makes them more likely to nurture that new relationship. So whether you thank a stranger for holding the

door or send a quick thank you note to that colleague who helped you on a project, acknowledging other people's contributions can lead to new opportunities.

2. It improves physical health. According to a 2013 study published in *Personality and Individual Differences*, grateful people experience less pain and report feeling healthier than the average. Unsurprisingly, grateful people are more inclined to take care of their health; they exercise more often and are more likely to have regular checkups with their doctors, which probably contributes to increased longevity.

3. Gratitude improves psychological health. It reduces a multitude of toxic emotions, ranging from envy and resentment to frustration and regret. Robert A. Emmons, a leading gratitude researcher, has conducted numerous studies on the link between gratitude and well-being. His research confirms that gratitude increases happiness and reduces depression.

4. It develops empathy and reduces aggression. Grateful people are more likely to behave prosocially, even when others behave less kindly, according to a 2012 study done at the University of Kentucky. Study participants who ranked higher on the gratitude scale were less likely to take revenge on others, even in the face of negative feedback. They showed more sensitivity and empathy toward other people, and exhibited a reduced desire for revenge.

5. Gratefulness improves sleep. According to a 2011 study published in *Applied Psychology: Health and Well-Being*, writing in a gratitude journal improves one's sleep. Spend just fifteen minutes writing down some grateful feelings before bed, and you may sleep better and longer.

6. It increases self-esteem. A 2014 study published in the *Journal of Applied Sport Psychology* found that gratitude increased the athlete's self-esteem, which is an essential component for optimal performance. Other studies have shown that gratitude reduces social comparisons and the tendency to compare oneself to others. Rather than feeling envious or resentful of those wealthier or in higher positions, one is able to appreciate their own accomplishments.

7. It decreases stress and develops resilience. For years, research has shown that gratitude not only reduces stress but can also play an important role in overcoming trauma. A 2006 study published in *Behavior Research and*

Therapy found that Vietnam War veterans with higher levels of gratitude experienced lower rates of PTSD symptoms. A 2003 study published in the *Journal of Personality and Social Psychology* showed that gratitude was crucial in overcoming the trauma following the September 11 terrorist attacks. Recognizing all that one can be grateful for, even during the worst times in life, improves resilience.

6   HAPPINESS

DANIEL LUMERA

Alice: "How long is forever?"
White Rabbit: "Sometimes it's just a second."
—*Alice in Wonderland* (2010)

The human being looks for happiness by following four paths:

1. The path of doing
2. The path of having
3. The path of appearing
4. The path of being

Authentic happiness belongs to the sphere of being.

EVERY DEATH IS A REBIRTH

It's September 2013, and my cell phone rings around 10:30 in the morning. His voice greets me and goes straight to the point, without preamble, direct and crude: "I'm dying. I was diagnosed with terminal pancreatic cancer and I would like to die in peace. Without regrets for the past and fear of the future. I'm calling you because I have a request. I want you to accompany me. If you accept, come as soon as possible because I have little time left. Maybe less than a month. I want to be alone, with you, without my children, without my wife. Alone. I need to forgive and meditate." There are phone calls that change your life. Unexpected. And they leave silence, to make room for listening. We're not afraid of death but rather of the eternity that awaits us. Enigmatic, unknowable, and infinite. That bottomless chasm that we carry inside. This is why many people anesthetize/numb themselves to life. They do everything possible not to hear it. To forget it. Yet acceptance of life on its own terms, despite death and pain, is the only gateway, the only real opportunity, to embrace life completely and really start living.

Infinity reminds us that everything is transient; nothing is permanent—that, ultimately, everything we have known and built will disappear, transforming itself again and again, until we no longer recognize the form of what now seems intimate and familiar. Every impression, every smile, and every moment will dissolve into eternity, as ephemeral as soap bubbles in a storm. Fear of death's darkness, loneliness, and coldness; the lack of a mother's embrace, or someone who can comfort us with even a word or smile. Everything will change. Everything is already changing, but we continue to cling with all of our might in order to restore balance, thus giving us a feeling of stability or safety even if it's an illusion. Instead, we should commit ourselves to finding stability in change, or really seek that part of us that is immutable and eternal as promised to us by the mystics and saints of every tradition. What is it that we fear—the end of everything? Do we fear the fact that death is the end of any contact with the people we love, of any relationship, of everything that warms the heart? The end—to no longer exist. And while we are afraid of ceasing to exist, we commit ourselves with all of our strength to not really living this life. Life goes on while we are committed to damaging our souls with toxic relationships, poisoned foods, or enslaving jobs, to pursuing dreams and goals without being aware that they come from frustrations and shortcomings. Are we free and happy? I mean, really happy? Are we free and happy right now? Because this moment is the only thing we really have. Death is a wake-up call. A ticking clock. The time has come. This is why we should live every moment as if it were the last one. Or the first one. With the same wonder, gratitude, amazement, joy, passion, fear, and love that small children have, or those who are consciously present in the last days and minutes of their life. Everything takes on depth, intensity, and urgency. This urgency is not haste but the understanding that these are the only moments, that one cannot lose the present. Hence the passion, intensity, and essentiality of those who are naked before themselves and life, of those who have understood, of those who open their hands and let go of all superfluousness, finally relieving their hearts of all those burdens that they themselves have accumulated and held.

In those moments, what really matters flourishes. Every look, action, thought, and emotion is truly forever. Everything is as if it were the last or first time. So, we should learn to live, to really live. This is one of the prerequisites for walking on the path of happiness, on the path of kindness. "Come. I want you to be the one who accompanies me to my death. I want to die in peace." Those words reverberate in me, especially said by someone who loves me, by a loved one. Who chose me. Sometimes the weakest child is called on to accompany a parent to their death. In those moments, unforeseen courage manifests at the right time, without getting lost in pain. I think of the instinct of a female dog who has just given birth to her puppies. She eats the placenta and cleanses the pups. There is no need for anyone to explain to her what she must do. We know how to give birth, but no one tells us how to die. Sometimes we need help to be born and die in peace and love. "Come." He had chosen me. I accepted. After a while I was on a plane. I do not remember anything of the trip but only his eyes on my arrival and the silence. We meditated in the silence and the light of presence, together. And together we forgave all of the past. Everything that had happened was heard. Every breath was a release. Every word, a song of gratitude. We were grateful for pain, grateful for love. Grateful for all that had been and was. Because gratitude, real gratitude, cleanses, liberates, opens the heart completely. To forgive one's whole life is to transform it into a gift, to understand its meaning and significance. It is embracing oneself in order to be reborn. With this spirit we can relive and face all our sorrows and enter the darkness—because the light is in there. Forgiveness and meditation. Love and silence. Silence and love. And every moment became forever. Together we reached the edge of time and we went beyond it, until "now" became "forever." Now is forever. When we really listen to ourselves, we don't need rules or someone to tell us when, how, and why to do things. We are like that dog. When now becomes always, then everything that happens is forever. It is a completely new sense of responsibility.

Everything, really everything that happens within ourselves in this very moment is the "forever" that scares us so much. It is a declaration of love, the most beautiful insight that a human being can welcome within themself. Now is forever. This moment is forever. That "forever" we have always sought in promises, in religions, in the universe, in love, in relationships, in children has always been here. We looked for it elsewhere but it was here. Right here. Hidden in this moment. It was always here. Now is forever. How would life change if we accepted that now is forever? That "forever" would become part of our sense of identity. We would act, feel, think, and choose through the lens of that "forever."

Time disappeared because we gave life to all of our pain, love, fear, and passion. That evening, in the silence, he looked at me. It made me feel small, fragile. And he said, "I've never been so happy." I really heard those words. Then I sincerely said, "Your body is dying. You leave two little children and a wife. You leave the things that made this life meaningful and successful, and yet you talk to me about happiness. Please teach me what kind of happiness this is because it doesn't depend on having health, wealth, success, fame, affection, or love. It doesn't depend on being able, going out, dancing free, going to visit your loved ones, traveling. Nor does it depend on being famous, recognized, or admired by others. What happiness is this?" Silence. And light. Light that shines in the eyes of those who are dying. He looked out the window, at the houses and city. These were his words as I remember them: "I have never been so present and awake in the miracle of life. I am so present, aware, and awake in this being, that a spontaneous existential happiness arises and blossoms in me for no other reason than the gratitude of being in this moment. No past or future. This is the greatest gift I have ever had in my life."

Doing, having, and appearing can create a sense of security, stability, strength, and satisfaction. The nature of happiness, however, is something else. We may choose, seek, love, or travel in the pursuit of happiness. But must we "do" or "have" something in order to be happy? Or is the nature of happiness inherent in our awareness of being? "A mule carrying a load of gold on its back doesn't know its value. Similarly, man is so busy carrying the weight of life on his shoulders, hoping to find some happiness at the end of the path, that he doesn't realize that he contains within himself the perennial and supreme bliss of the soul. Since he seeks happiness in things, he doesn't know that he already possesses in himself a treasure of happiness." These words, by Paramahansa Yogananda, come to my mind. He was the first Eastern wise man to arrive in the United States in the first half of the twentieth century.

Just think what a revolution it would be if our choices and actions were motivated and pushed not by the search of happiness but instead as its expression. I choose because I am happy and not in order to be happy. How much would the quality of your life, relationships, and work change?

Utopia? One thing is certain. Each of us will deal with death. And perhaps the lucky ones will learn to experience this kind of happiness at that moment. But why wait? Why not take that step now? Shake off heaviness, inertia, indolence, ignorance, greed, jealousy, and resentment. Stop anesthetizing yourself with food, sex, toxic relationships, television, and media. And wake up. Here. Now.

He died happy.

THE KINDNESS OF THE HEART

He didn't call himself master but he was. I've never seen him angry. There was no trace of sadness, loneliness, or arrogance in him. Balanced in life, he sat in the chair of simplicity. Anthony Elenjimittam, one of Mohandas Karamchand Gandhi's last disciples, lived his last days in Assisi, Italy. He was born and spent the first part of his life in India, where he met the Mahatma and received an ecumenical mandate from him. People called Elenjimittam father because of his past in the Dominican Order, a Catholic mendicant order, which ended when he decided to embrace an interfaith vision. He was ninety-six years old when he left this earth. He lived alone and ate only what people who loved him brought him from time to time. A year before his physical death, he passed out and was found several hours later. He narrowly escaped death. When he woke up, the first thing he did was smile. He simply said, "I was doing dress rehearsals." His mind was immersed in a high state of consciousness that allowed him to see beauty as well as live in a state of joy and happiness. But above all, I remember his kindness. His heart surrendered to kindness. And as such, he gave kindness to anyone who approached him. When he spoke, you felt there was no presumption of being right or wanting to impose a truth. It was just sharing. A form of kindness from those who have lived in the universe, and have love in their heart and give it as a gift. Because he chose not to do otherwise, longevity was combined with incomparable mental freshness and juvenility. He spoke eleven languages, including Latin and Sanskrit. He used to say that the secret of a forever young mind is the result of the foods we feed it. He nourished his mind through the silence of meditation, happiness and joy of awareness, purity of devotion, and integrity of discipline. With him, one could study the thousand-year-old texts of the Indovedic tradition (such as the *Sūtra of Patañjali* and *Upanishads*) and find the same teachings in the texts of Saints Augustine, Teresa, and Francis. He built bridges and broke down the walls of mind and consciousness. Several times he said, "I want to stay where the truth lives, wherever it is and in whatever form it manifests itself." Kindness is something we can give and receive. When this happens, when we find someone willing to give this gift, the heart becomes light and smiles because it remembers how it should live.

Father Anthony's mind and spirit was like that of a child: a state of enthusiasm in which everything was a discovery, always seeing and doing and saying things for the first time. Even though he repeated some concepts for over fifty years, he always lived them with a new spirit every time. His mind had found the fountain of youth. Everything was new to him. In this way, he lived, loved, and felt life. In addition to enthusiasm, the side effects of such a pure and clear mind are also wonder, kindness, inner silence, equanimity, compassion, and love. The core of his lifestyle was his meditation practice. His mind was forged in contemplative discipline, and his inner gaze was constantly fixed on the infinity and its vastness. And it's thanks to this vastness that we become simple, essential, sober, and humble—the humility that resides in kind hearts that truly know what love is.

HOW DOES ONE ACHIEVE THIS KIND OF HAPPINESS AND MENTAL YOUTH?

It's actually much simpler than you think. Many traditions (for example, Buddhism, Zen philosophy, Taoism, and Vedantism) have studied the state of the mind and states of consciousness, and taught the art of meditation, contemplation, inner silence, presence, and gnostic as fundamental tools to achieve, understand, and realize the nature of true happiness. Happiness is a natural state in the conscious individual. It doesn't depend on doing, having, or appearing but instead only and exclusively on the awareness of being: awake and totally present in the miracle of life. When the mind becomes silent, a clear and intense awareness of being (beyond all forms, judgments, and levels of identification) shines on it so as to foster spontaneous happiness—an an intrinsic characteristic of awareness.

In the chapter dedicated to the perfect mind, we will see how it's possible to understand the nature of the mind, what influences it, and how to purify it in order to directly experience the intrinsic happiness of life. Ancient traditions have pointed to both interior and exterior lifestyles in order to achieve a better quality of relationships, nutrition, health, mind purification, and vital energy awakening along with the correct use of the mind and meditation.

Today, science is validating with objective data what thousands of years ago was part of a corpus of teachings at the base of a master way of well-being and self-realization. Modern science considers three pillars of well-being: healthy eating, physical activity, and meditation. Since the human being is an integrated system of multiple levels (mental, emotional, and physical) that is influenced by the external environment and in turn influences the external environment, an integrated approach is extremely effective. The nature of happiness is achievable through a path of kindness that considers all aspects of life, celebrating them in a new light that also includes lifestyle, nutrition, physical movement, the quality of relationships, the inner environment, mindfulness, compassion, optimism, forgiveness, and meditation.

HOW THEY SELL US HAPPINESS

This society is made up of an economic system that follows a simple rule: obtaining the maximum benefit at the lowest cost possible. The *Homo economicus* is defined as "rational," in that they maximize their well-being and personal gain as an objective. The mathematical function defining this process is called the "utility function." For this reason, purchasing radioactive flour (with all the adverse consequences) rather than natural organic flour is justifiable, if it justifies one's own self-interest and need. We all buy a little "radioactive flour" every day. This rationality, lacking any feeling and value, follows three simple steps: being clear about one's preferences and interests, maximizing one's satisfaction by making the best use of resources, and analyzing as well as predicting the situation in order to make the most convenient choice. When this type of rationality mixes with *Homo sapiens'* frustrated ego, the consequence is visible to all—starting from the environment and nature.

Entire lives are dedicated to working in an inhumane way to produce things that others, working inhumanely, will buy. Did all of this ever bring real and lasting happiness, and did it ever give deep meaning and purpose to these people's existence? It's an economic system, the current one, which transforms the pursuit of happiness into a productive prostitution based on numbers. In the age of profits, the main purpose is to increase numbers, to the detriment of everything else, including human life, other beings, and nature.

In 1972, the fourth king of Bhutan, Jigme Singye Wangchuck, was still a teenager. During an interview about the country's gross domestic product (GDP), he replied that the only important thing was gross domestic happiness (GDH). From that moment on, Bhutan jumped into the international limelight mainly for its GDH index. Long before the GDH was born, Bhutan was influenced by Buddhist culture, which focuses on the awareness and realization of human happiness and harmony with the environment, and by extension, the well-being of the population. Since 1972, Bhutan has dedicated a large space to happiness, so much so that in 2008, the fifth king, Jigme Khesar Namgyel Wangchuck, included the GDH in the first democratic constitution, defining it according to four criteria: protection of the environment, defense of local cultures, good administration, and sustainable development. It's a development policy based on profound values. All the laws and proposed projects pass through the GDH Commission, which ensures its compatibility with the policy of happiness and can prevent their implementation even at the expense of substantial financial revenues. For example, if a project has a highly negative environmental impact, the commission can reject it in order to protect the real well-being of the population, which is its priority. The socioeconomic balance is thus not altered by the pursuit of mere profit.

The GDH is difficult to calculate because it tries to objectively measure emotions such as jealousy, fear, security, and joy—all hardly quantifiable feelings. Therefore, every five years, the administration conducts surveys through questionnaires given to a representative sample of the Bhutanese population. Questions like, "Are you jealous of your neighbor?" "Do you respect wildlife?" and "What's your relationship with money?" The results make it possible to note improvements or deterioration, and based on the results of the survey, recalibrate policy actions. Paradoxically, the goal is not to make people happy but rather to create personal and environmental conditions that allow anyone who wants it to be happy.

The concept of happiness here is not the same as in the West; it's not a momentary feeling of satisfaction and pleasure for having received or obtained something gratifying but instead something much deeper. A state in which one experiences total contentment. In the West, happiness is intended as a degree of satisfaction (life satisfaction), fulfillment, pleasure, and the subjective perception of sensory well-being. Happiness is satisfying sensory pleasure through doing or having. The truth is, happiness marketing sells substitutes for well-being, euphoria, and transient sensory satisfaction. Happiness has become the illusion and promise of a favorable destiny in terms of material wealth. From a natural state of full awareness, integrity, fullness, joy, and consciousness of the miracle of life, happiness has been downgraded to a goal to be achieved based on the capacity for sensory and material satisfaction. The current economic system generates fake and superfluous needs. We allow an illusionary idea of happiness to be sold to us because we don't know the real nature, and believe it depends on doing, having, and appearing, and not on the awareness of being. One of the hardest myths to bust is the idea of happiness being linked to money. Daniel Kahneman, winners of the 2002 Nobel Prize in Economics, introduced the concept of a national well-being audit, demonstrating that material well-being has little impact on the real perception of happiness. Having more money doesn't increase the "moment-to-moment happiness," the individual moments of happiness. Money as the source of happiness is false for the simple reason that it creates short-lived fulfillment, offering a feeling of security, satisfaction, and euphoria, but not real happiness.

A 2020 meta-analysis of over five thousand participants from the Higher School of Economics and University of Toronto demonstrated that the brain's reward center is activated by sex, good food, and the "money stimulus," and that all three are processed by the same area of the brain: the basal ganglion. It's sufficient to fantasize about a large cash win to immediately feel security, contentment, consolation, and gratification. All of this cannot be called happiness in any way; on the contrary, money is linked to competitive stimuli. Glenn Firebaugh and Laura M. Tach, two sociologists at Pennsylvania State University and Harvard University, respectively, set up a social experiment in which they asked participants to choose between two possible job options. Option one was an $80,000 annual salary for a position where, for the same job, one's coworkers made $90,000. Option two was an annual salary of $60,000 for a position where, for equal work, one's coworkers received $50,000. What do you imagine most of the participants chose? The second option. We prefer the feeling of superiority to economic advantage. It's competition that pushes us to feel superior. This insatiable greed leads us to feel dissatisfaction and frustration if someone earns, has, does, or appears more and better than us. This competitive race originates from a deep sense of dissatisfaction that can't be cured through accumulation or infinite growth. There will be no doing, having, or appearing that can give us real and complete satisfaction. Ancient knowledge, embedded deep in our souls, tells us that the answer lies in the awareness of being: the origin of true and lasting happiness.

THE STAGES OF UNDERSTANDING HAPPINESS

Tell me what your happiness is and I will tell you who you are. Every human being can understand and experience happiness based on their level of awareness. If we were to categorize distinct levels of understanding, they would be subdivided like this.

The first level can be defined as "instinctual" because the human being is dominated by passions, senses, and emotions. The purpose of life is fulfilling sensory pleasures. The mind co-opts other people's ideas and thoughts. The sense of right and wrong is established on the basis of what is commonly considered correct or incorrect, without any personal discernment. The individual simply adjusts to what is presented as right or wrong without questioning it. In this stage, failure and success are associated with pleasure and pain. The concept and experience of happiness have a purely sensory dimension, satisfying and enjoying passions and instincts.

In the second level, the concept of happiness is explored through the logic of convenience. The human being becomes a "merchant of happiness," seeking a personal advantage in everything. Here happiness is exchanged for satisfying benefit, interest, and personal growth. Right and wrong are established on the basis of the questions: "What advantages do I have?" "What do I get from it?" In this phase, we believe that reaching happiness depends on how much we are able to satisfy our personal interests.

In the third level, however, the human being explores the experience of happiness through giving and helping others. When this phase is underdeveloped, "selfish altruism" manifests itself—that is, doing good to others is useful in order to obtain admiration, approval, love, and success. We are still in a merchant phase because there is more profit and convenience in giving than in accepting. Yet even in this first underdeveloped aspect, the interests of the individual begin to include those of others as well. Happiness is being admired and loved, accepted and valued. When this stage reaches maturity, one feels the need to expand the idea and perception of oneself, including others: one's happiness includes others' happiness and satisfying one's needs includes satisfying others' needs. The desire to be authentically useful arises spontaneously. It's here that human beings seriously consider the possibility of sacrificing their resources, time, and talent in service of other people's good. The concept of right and wrong also radically changes, because feelings of love, compassion, and selfless service for the good of others no longer depends on personal satisfaction. In its highest phase, the human being conceives the possibility of truly putting one's life at the service of others. Happiness is experienced through giving oneself and serving others' happiness.

In the fourth and final level, happiness is no longer attributed to the sphere of doing, having, appearing, and giving, but it's experienced through the awareness of being. To a human being going through the previous stages, this level of awareness is often incomprehensible. In this dimension, we experience the desire to be an expression of an impersonal and universal conscience: a pure existential conscience. Happiness is experienced as a natural state of consciousness: an inherent and natural trait of the pure awareness of being. In this level of consciousness, the perception of being one with every form of life manifests itself, and the ultimate purpose of existence becomes realizing and sharing this state of consciousness. You no longer need to do anything or have anything to be fully and authentically happy, no longer needing definitions, forms, actions, possessions, and merits to define yourself.

THE PURSUIT OF HAPPINESS

Are you happy? How many times have we heard or asked this question? "Yes, my job/relationship/family is going well. I just got back from a wonderful trip" or "No, this is not a good time because of my health." What kind of happiness are we talking about?

All the great masters of every tradition, from the Athenian philosophers in the West to Confucius and Lao-tzu in the East, to the most ancient Indian millenary wisdoms, have questioned themselves about the nature of happiness. As the centuries have passed, two philosophies, perspectives, and paradigms have clearly arisen, giving way to many empirical investigations in the last century.

The first is hedonism, which is based on the concept that happiness comes from the maximization of pleasure and personal satisfaction. The second major trend is so-called eudaimonic happiness, which means happiness as the realization of one's authentic potential as a human being. These two schools of thought have asked different questions about the relationship between human beings and happiness, exploring often complementary areas, and prescribing distinct ingredients and recipes.

THE WAY TO HEDONISTIC HAPPINESS

Viewing well-being as pleasure and hedonistic happiness has a long history. The first Greek philosopher who wrote about it in the fourth century BC was Aristippus, who taught that the goal of life is to experience maximum pleasure and that happiness is nothing more than the sum of moments of pleasure. Philosophical hedonism has been followed by many others, such as Epicurus, who declared pleasure to be the only intrinsic good; Thomas Hobbes, who believed that happiness resides in the satisfaction of human appetites; and the Marquis de Sade, convinced that the pursuit of pleasure was the ultimate goal of life.

Thus, toward the end of the 1890s, hedonistic happiness caught the attention of psychologists, who considered personal satisfaction and subjective well-being as the main indicator for evaluating and measuring human happiness. This focus by psychologists significantly influenced collective perception and human behavior. There are three main ingredients to this formula: the presence of positive emotions, absence of negative emotions, and maximum satisfaction in life. This movement had a profound impact on the Western world in the last century. It tipped the balance from a concept of happiness belonging to the being, imbued with the fundamental values of the human being, to the mere search for one pleasure after another. This gave life to what may seem one of the greatest paradoxes of the modern era: the search for hedonistic happiness as a state of chronic dissatisfaction.

In 1974, a US demographer and professor of economics, Richard Easterlin, realized that something was wrong, and began to collect an immense amount of data on the relationship between per capita wealth and subjective well-being. He noticed that the countries with per capita GDP below the poverty line were also those with the lowest level of subjective well-being. So far so good, it makes sense. Following studies contradicted this idea, giving rise to what is now known as the Easterlin paradox or paradox of happiness. This paradox was analyzed by many scholars around the world, including Andrew Clark, who in 2008 published a surprising result in the *Journal of Economic Literature*: as income grows above a certain threshold, the correlation with the increase in happiness weakens, until it disappears completely and in some cases is even reversed. In 2013, Eugenio Proto and Aldo Rustichini's research revealed even more: in countries with a per capita GDP of more than $30,000, the higher the wealth, the lower the happiness intended as hedonistic happiness.

Of course, such a sensational and apparently counterintuitive finding resulted in years of intense debate, still ongoing among scholars of every discipline. Gradually, this paradox went beyond the money factor and extended to all of those external factors that we consider responsible for well-being. The concept of wealth itself has also been expanded to include new categories such as relational goods (including the family and emotional sphere as well as participation in the social and political life of one's community) and environmental assets—commodities that money can't buy, frequently sacrificed in order to get the income necessary to purchase consumer goods.

THE TREADMILL EFFECT

But is decreasing happiness with an exponential increase in available money really such a counterintuitive result? Let's go back for a moment to Darwin's theory, which we mentioned in the preface, and in particular the postulate that "it's not the strongest or most intelligent species that survives, but the one that better adapts to change." Easterlin's paradox covers hedonistic adaptation, also known as the *treadmill effect*. This concept, widely tested by psychologists, refers to the general tendency of returning to the same level of happiness after both a grand "good" and grand "bad" life event. Whether it's a lottery win, marriage, or new job position, or divorce, job loss, or health problem, after an initial peak of happiness or sadness we adapt, and the impact of the event in the medium term and long run becomes minimal. In perfect synchrony with our adaptation instinct, to survive as a species, circumstances have little impact on the quality of our life and happiness—10 percent to be precise. Yet the system we live in is based precisely on a longing for external palliatives to give us temporary pleasure. It's like a nonstop run on life's great treadmill, only to realize we're in the exact same place. Where are we running? What are we running from? What are we looking for?

THE PLEASURE BUBBLE 3.0

A phrase that has been successful in social networks goes something like this: "Children learn a fundamental lesson very early and they learn it by playing video games; the lesson is that nothing will kill them faster than standing still." If you've ever played, or watched your child or grandchild play, a video game, you may have noticed that when you stop, something happens, and adversaries immediately start chasing you, outrun you, and put you at a disadvantage or even kill you. Something "bad" always happens when you stop. Children learn precisely through this dynamic that it's necessary to move to survive, to be in a constant frenzy, ready for the next move, whatever it is. While this is inspiring because it teaches us how to flow through all happenings, it's also limiting. I'm not talking about teaching children or adults to stop, which can be associated with a feeling of immobility, castration, or inability to express themselves, but just to be still.

Being still is something meditators know very well; it's that moment when the whole world stops because you're still. That instant in which you stop and let the world flow in front of you, without the need to respond to its dynamics, without defining or judging it, but simply observing it from a detached, transcendental point of view. This gives you an enormous advantage, being steadfast while the hurricane swirls around you, observing without being part of everything that normally absorbs you, drags you, makes you suffer, and destabilizes you. You can look at, observe, listen to, and understand it from a completely new perspective, and through it, see, hear, and understand yourself. In being still, you realize how many times you are devoured by doing, by believing that to fulfill yourself and be happy, you must acquire appearances, possessions, and always do-do-do because it's the only way you will feel good. This is a myth that needs to be debunked in order to rediscover our natural condition of real well-being. Being still is an inner state where you sit down, close your eyes, breathe deeply, and just listen. You listen to everything that happens inside and outside you, no longer adhering to it or reacting to those stimuli. Five minutes of practice every day should be enough to develop this inner state of silence along with an attention to the breath and presence. Day after day, it's easy to notice how this skill manifests itself, flourishes, and allows you to consciously direct your life's design. There are times when it's not necessary to stop but instead be still so as to observe ourselves and life, and finally really see each other.

THE WAY TO EUDAIMONIC HAPPINESS

The Greeks called happiness *eudaimonìa*, literally the condition of a "good spirit" (*eu*, "good," and *daimon*, "spirit"), one who is possessed by a good demon, by a good fortune that allows them to prosper; the effect is a happy, positive, and permanently pleasant tone of the soul. In short, we would say that it's being born under a lucky star.

For the Greeks, happiness was a state of mind in and of itself, and not related to a particular event that sparked it. The good "demon" watched over people and the polis, and gave happiness. The concept of *eudaimonia* was created by Aristotle, who referring to the *daimon* intended as "true nature," believed that happiness was a crass idea, emphasizing that not all desires deserve to be pursued. Even if some of these desires could give pleasure, they would not produce well-being. In the *Nicomachean Ethics*, Aristotle states that the highest of all goods obtainable by human actions is "eudaimonia." The Aristotelian model focuses on the virtuous individual and those traits that qualify them as an expression of values. Furthermore, in this model, the virtues of the soul consist of two types: virtues of thought and character. The first type is born and grows mainly from teaching; that's why it needs experience and time. The second type comes instead from habits and lifestyle choices. In the Aristotelian model, "growth" or "change" becomes a fundamental dimension; the individual is understood as constantly being pushed forward by a dynamic principle toward what is better or perfect. The individual is thus seen as aspiring to positive goals and values, and striving to achieve them. In this model of happiness, pleasure is traced back to what one feels when training talents and virtues, and increases the more the ability is realized or the greater the complexity. Carol D. Ryff and Burton Singer's research, pioneering the vein of eudaimonic happiness, explored it by giving a new definition of well-being as "effort towards perfection represented by the realization of one's true potential." They introduced a multidimensional approach that touches on six distinct aspects of human actualization: autonomy, personal growth, self-acceptance, life purpose, mastery, and positive relationships with others. This new approach made it possible to shift attention from well-being mainly defined in terms of pursuing pleasure to happiness understood in a broader sense than human experience.

PLEASURE OR HAPPINESS?

THE MARSHMALLOW EXPERIMENT

> It's a story about us, people, being persuaded to spend money we don't have on things we don't need to create impressions that won't last on people we don't care about.
> —Tim Jackson, TED Talk 2010

The "marshmallow experiment" is one of the most famous and significant in the field of psychology. Started in the 1960s by a young Stanford psychology professor named Walter Mischel, the experiment is simple, involving a group of children between three and five years old. Each child is left alone in a room with a quantity of marshmallows. Before the researcher leaves the room for about ten minutes, they tell the child that they're free to eat whatever they want, but if they resist the temptation to eat immediately, they will receive more.

The results were clear: a third of the children ate the sweets immediately, the second third did not wait for the required time and ate the sweets within six minutes, while only the remaining third of the children waited long enough to have more sweets. These results may not be surprising because for children, those minutes are endless and they find it difficult to resist the candies.

What made this experiment one of the most famous in the world is that these children were followed over time into adulthood, and for forty years they were observed from the perspective of character, career, satisfaction, and happiness in their lives. Guess who had the best outcomes? Guess who had excellent results in school, high scores on university tests, a successful career, an optimistic and problem-solving style of thinking, and a significantly above average level of happiness? Exactly those children who have been able to postpone pleasure in view of a more important goal.

But that's not all, because this experiment has been repeated many times, revealing additional results: subjects also showed higher self-esteem, a better way of managing emotions, and less predisposition to drug abuse. On the other hand, those children who could not wait showed a tendency as adults to become overweight or obese, and generally presented with worse health. In this experiment, it's not difficult to see ourselves trying to get fit but then giving in to the first dessert, or when we want to pursue an important project, instead we find ourselves compulsively scrolling through social networks, or when we choose to start meditating, a bad day is enough to break our resolve.

If we further broaden the scope of these discoveries by applying them to today's reality of "everything and immediately," perhaps it's time to ask ourselves, Where are we going? How are we going? And what's the meaning we really want to give to our lives? Do we want to pursue pleasures made up of a series of immediate, external, and temporary gratifications, or do we want to rediscover a deeper happiness that has its roots in the values of patience, determination, self-listening, and real awareness? It's not just about self-control but also about making a conscious choice that reminds us of the immense potential that is in each of us.

THE HIDDEN COSTS OF PLEASURE AND GENETIC RESEARCH

In a 2013 study, Barbara Fredrickson and colleagues at the University of California at Los Angeles examined the biological influence of hedonistic and eudaimonic well-being on the human genome. Interested in the pattern of gene expression within people's immune cells, they found surprising results. Although both types of well-being were previously associated with better physical and mental health, this new research found that eudaimonic well-being is associated, at the molecular level, with a significantly reduced expression of the stress-related CTRA gene along with an improved immune response and reduced inflammation at the cellular level. Conversely, hedonistic well-being yielded opposite results: the association was a diminished immune response and increased tendency to inflammation. But there's more. Although the results of the two groups were opposite on a biological level, people in both groups stated that they felt a greater well-being at the end of the experiment. One possibility is that people oriented toward a life of "pleasure only" consume the emotional equivalent of those empty-calorie foods that provide no real nutrition; everyday activities pursuing hedonism are comparable to foods such as sweetened drinks and refined sugar, which provide short-term happiness, but have long-term negative health consequences of which we are not immediately aware.

Small pleasures therefore give us the impression of happiness, but do not help us broaden our awareness or realize our talents, skills, and abilities. This is an "evolutionary disadvantage" that seems to be directly at the cellular level. The body doesn't lie and responds better to a different kind of well-being, based on a sense of connectedness and purpose. This can reveal the hidden costs of purely hedonistic well-being as well as open up new perspectives for studies that directly link the search for life's purpose to its quality and length.

At first glance it seems clear that the eudaimonic way, the one that pursues a more satisfying and lasting happiness, is better and more convenient from an evolutionary point of view. There is, however, a third way to happiness.

Is it possible to live in a state of happiness regardless of what happens, whether a great love, great disappointment, or great pain? Can one be happy regardless of what one chooses to pursue, do, have, or achieve in life? It happens in an instant, the same moment in which we realize that we are alive, awake in full awareness of existing. Happiness manifests itself as a natural state of being when you are fully aware that you exist—an intense, unconditional, existential happiness. Thus, every previous path fades, and past and future merge, disappearing in the only moment that really exists, in the infinite possibilities of this present that opens up like a bud in spring: the miracle of life. We should remember to celebrate this miracle with every breath, and with the same intensity as the last moment.

What if every thought, word, action, pain, love, and loss were simply the expression of the infinite happiness that exists regardless, inherent in every moment? No longer something to seek, conquer, or fight for, but something that belongs to the very essence of the human being, experienced through a sense of conscious identity: the celebration of eternity in us, the existential happiness inherent in us. This is the real revolution. We are doomed to happiness; we have to surrender to it.

7   THE SCIENCE OF HAPPINESS

IMMACULATA DE VIVO

Whoever is happy will make others happy too.
—Anne Frank, *The Diary of a Young Girl*

HAPPINESS, THE ENDLESS SEARCH

Can happiness be the object of scientific studies? Are we able to measure it, translate it into numbers, and treat it as a mathematical datum? Scientists' efforts in the last decades have often gone in this direction, sometimes supplying important results that have led to considerable progress in the knowledge of this central aspect of life.

According to the World Health Organization, happiness is one of the fundamental components of human health and involves a great variety of disciplines, from medicine to economics, from psychology to neuroscience to evolutionary biology. Science has studied the psychological consequences of many different happenings in life, such as lottery winnings, political elections, economic fortunes, job losses, socioeconomic inequalities, divorces, illnesses, and mourning. Many aspects have been highlighted, while others still remain to be clarified, but science as a whole has managed to set some constants that allow us to identify the fundamental characteristics of happiness and its effects on well-being in order to know it better as well as get as close to it as possible.

HAPPINESS IS COLLECTIVE, HAPPINESS IS CONTAGIOUS

One of the most interesting studies on this matter appeared in 2008 in the prestigious *British Medical Journal,* and was conducted by Nicholas

Christakis from Harvard Medical School and James Fowler from the University of California at San Diego. They analyzed the dynamics through which happiness propagates within a vast network of individuals. The pair started from the assumption that happiness should not be considered only at the individual level but rather as a collective phenomenon, influencing people who are connected to each other.

Based on previous studies, the two scientists highlighted how emotional states in general, not just happiness, can be transferred directly from one individual to another through mechanisms such as imitation or "emotional contagion." People tend to reproduce body movements and facial expressions with significant emotional content that they see in others, appropriating them together with the feeling that generated them. Emotional states "acquired" by observing others can persist for a few seconds, but also for several weeks. Students who were randomly assigned to share a room with a mildly depressed roommate showed increasing signs of depression over three months.

Other dynamics of emotional contagion, even in situations of fleeting contact, have been observed in experiments where the table service staff of a restaurant, invited to be open toward and smile at customers, received greater satisfaction feedback and larger tips.

Christakis and Fowler wanted to focus their attention on the contagion of happiness within a complex social network to understand if the emotion transfer between subjects linked by stable relationships persists longer over time and travels farther, passing from direct ties to indirect ones. To do so, the two scholars observed a cohort of subjects for twenty years participating in the Framingham Heart Study, a famous and important epidemiological study begun in 1948 in the town of Framingham, Massachusetts, to analyze the mechanisms of cardiovascular disease within a community.

The study on happiness involved 4,739 individuals, followed from 1983 to 2003, whose happiness levels were periodically measured through questionnaires. The data were processed by software and displayed in a graphic image in which each node represented an individual, each line represented a link, and different colors represented the various degrees of happiness. Researchers were able to conclude that happiness tends to radiate from a happy individual up to three degrees of separation (for example, a friend of a friend of a friend of ours), thus creating clusters in which people in contact with happier people are more likely to become happy themselves in the future.

The longitudinal nature of the study, which lasted twenty years, made it possible to understand that these variations don't depend on people tending to associate themselves with similar personalities but instead on a true "propagation of happiness" from one individual to another through the network of relationships they are part of. The intensity of this "contagion" depends on the role that each individual plays in the life of another along with the direct or indirect contact between them. The effect tends to decay over time or with increasing geographic distance.

The conclusion of this fundamental study is that the happiness of a single individual, although conditioned by a large number of personal variables, depends to a large extent on the happiness of the other individuals to whom they're connected and therefore is a collective phenomenon.

HAPPINESS AND HEALTH

Happiness is still being studied with respect to its interactions with health and illness, but there already are studies that link this emotional state to a longer life expectancy. This is the case in a 2015 research study conducted by the Universities of Colorado and North Carolina that cross-referenced data relating to levels of happiness with mortality rates. It found that compared to people identified as "very happy," those "fairly happy" have a 6 percent higher risk of death, while the "unhappy" ones have a 14 percent risk of death, regardless of variables such as marital status, socioeconomic level, or religious confession.

Investigating the causes of happiness, scientists have identified some particularly relevant factors: character traits and individual behavior, emotional status, level of education, religious faith, physical health, and sexual activity. Now the focus has shifted to the consequences of happiness, especially on physical and mental health. Studies conducted on smaller but interesting samples showed that happiness, measured through the analysis of the language used to describe some events in one's life, corresponded to a longer life expectancy compared to those who used gloomy tones and had a negative view. Happiness appears to be a factor inversely proportional to perceived stress and could help protect against disease, thereby improving the immune response. Happy people tend to have better health because they have greater adaptability, stronger problem-solving skills, and more effective resistance strategies. They appear more creative, have a strong imagination, and are better at integrating information and experience. They show resilience and greater ability to cope with adversity. This research needs to be confirmed in larger samples, allowing us to generalize the results. But these findings already illustrate the protective potential that happiness has on our health.

In more recent years, science has tried to limit the broad and multifaceted concept of happiness in an attempt to make it observable as well as measurable according to current research methods. There is no scientifically shared definition at the moment, and the efforts of some researchers have concentrated on identifying the various "dimensions of happiness" so that they can be analyzed as subcategories of the broader concept to which they belong.

Laura Kubzansky is codirector of the Lee Kum Sheung Center for Health and Happiness at the Harvard T. H. Chan School of Public Health, a Harvard University institute dedicated to studies on happiness as a tool for promoting health. She's also my friend and neighbor; together we've conducted collaborative research and had the opportunity to discuss her fascinating studies. Her research center has decided to focus attention on those psychological dispositions that contribute to determining happiness—or unhappiness—and are more easily observable, such as optimism, which we have already talked about, and a sense of purpose.

On the latter topic, Kubzansky published a report in 2019 looking at the relationship between a strong sense of purpose in life and the risk of cardiovascular disease. Being strongly motivated is linked to a reduced risk of developing this kind of disease—a beneficial effect that unfolds through three different mechanisms: strengthening other psychological and social resources that protect against the the cardiotoxic effects of excessive stress (not perceiving some stressors as such or responding to them in an emotionally moderate way), adopting healthy behaviors (indirect effect), and activating biological processes (direct effect, such as lowering chronic inflammation).

In 2013, a study by the University of Michigan analyzed a sample of 6,739 adults older than fifty, noting that subjects with higher levels of motivation and sense of purpose also recorded significantly lower stroke risk rates. The effect of a sense of purpose as a component of happiness on physical health has yielded several interesting findings. In 2019, the same university studied an additional sample of 6,985 individuals, detecting a significantly lower mortality rate in people with a strong motivation.

In 2017, Kubzansky published a study in which she investigated the relationship between having a purpose in life and the state of physical functions in older people. In particular, she wanted to verify the correlation with two functions used in medicine as markers of aging and frailty: the ability to grip objects with the hands and one's walking speed. The study involved 4,486 people of both sexes over the age of fifty whose sense of purpose in life was measured and quantified through a value scale. After four years of observation, it was seen that an increase of one point on this "purpose scale" corresponded to a 13 percent reduction in the risk of developing poor grip skills and a 14 percent reduction in the risk of slow walking. Having a purpose in life, feeling motivated, and pursuing a goal is a modifiable factor that we can intervene in so as to encourage positive action toward our health.

TAKING CARE OF YOUR WELL-BEING

Andrew Steptoe, from the University College of London, highlighted recent scientific interest in investigating the decline in happiness not only as a consequence of the onset of a disease but also as a possible cause of the disease or rather as a risk factor. Many studies have linked happiness to reduced mortality and more disease-free years. Being satisfied with your life motivates you to take care of yourself, even in simple daily tasks such as exercising, eating fruit, using sunscreen, and avoiding smoking. It was even possible to observe a correlation between happiness and the levels of fat around the heart, with patterns of raising and lowering cortisol levels among unhappy people compared to those who declared a greater state of well-being.

Scientists are rethinking the role of science and medicine, no longer considered just as tools against disease, but as a support for well-being practices even for healthy people.

According to Kubzansky, the ironic aspect of the term *health care* is that it actually has little to do with health and a lot to do with disease because it focuses more on curing the pathology than creating stable conditions of well-being that prevent the disease itself. But health is much more than the absence of disease; just as this sets in motion a series of physiological reactions, happiness too, with its components of optimism, motivation, and a sense of purpose, can influence our body functions. That's why Harvard University established the Lee Center for Health and Happiness, in line with Seligman's idea that psychology should contribute to increasing people's awareness of their own well-being. Science can help us not only when we are sick but also and above all when we are healthy and aspire to improve our mind as well as body's well-being.

II INSTRUMENTS

8   THE QUALITY OF RELATIONSHIPS

DANIEL LUMERA

A well-tied knot needs no rope and yet none can untie it.
—Lao Tzu, quoted in *The Way of Lao Tzu (Tao-te Ching)*, translated by Wing-tsit Chan

The quality of relationships is one of the six pillars for a happy, healthy, and long life. Relationships are real food, and as such they can be toxic or rich in emotional, vital, and mental nutrients. Life itself is a constant relationship, not only with one's own kind, but with animals, plants, objects, ideas, and abstract concepts.

THE IMPORTANCE OF RELATIONSHIPS IN THE NEAR FUTURE

In the near future, the ability to develop healthy, happy, and fulfilling relationships among human beings will be a personal and social skill of fundamental importance for the quality of life as well as survival on the planet itself. According to UN estimates (updated to 2019), in 2011 the world human population reached the threshold of 7 billion, just twelve years after reaching the 6 billion threshold, and it is estimated that in October 2019, it had reached around 7.7 billion. Forecasts say that we will reach about 8.9 billion *Homo sapiens* in 2035, about 9.8 billion in 2050, and 11.2 billion in 2100. According to data provided by the Central Intelligence Agency in 2007, the growth trend was 211,090 people per day; this means a potential 80 million people more each year who will compete for food, water, and resources. Most of the growth takes place in underdeveloped areas, where the quality of life is far below standard; almost 2 billion people live on less than a dollar a day, and many of them don't have access to clean water and sufficient food. The more the number of human beings on the planet increases, the more significant the relational

(continued)

component becomes; relational skills are the basis of peaceful coexistence and a better quality of life.

1830: 1 billion *Homo sapiens*

1930: 2 billion (doubled in a hundred years)

1975: 4 billion (doubled in forty-five years)

1999: 6 billion

2011: 7 billion (almost doubled in twenty-five years)

2050: 10 billion

Growth trend: over 200,000 new individuals per day

LOVE IN RELATIONSHIPS

Relationships can become a means to experience love. The most common expression of relational love is undoubtedly the passionate kind. It's a type of experience that seeks love through emotional intensity, strong physical attraction, sensory drives, and instincts. The more you are mine, the more it means that I love you. Possession, belonging, and physical passion are the credentials to demonstrate as well as experience love. In a slightly more evolved phase, instinct gives way to sentiment, leaving more room for the partner's inner characteristics and virtues. The relational exchange shifts the focus from mere physicality to personal feelings and connections, based on a mental state of pleasure in being together along with mutual help and support in achieving one's goals in life. This is certainly a more serene relationship than a physical one.

In the experience of love in relationships, there is also a much rarer and different aspect in which attention is entirely directed to the interior. The characteristics of consciousness and the spiritual ones are brought to the foreground. The couple experiences love through sharing high purposes, and exploring and celebrating the spiritual plane of existence. In this context, spirituality doesn't refer to religion but instead to the meaning and purpose of existence in its universal sense. Among all of these types of relationships, there is clearly a range of expressions, and a couple may experience them all at the same time, to different extents depending on the moment. The human being, trying to experience and find love in relationships, makes different attempts.

EXPERIENCING LOVE

Experience love through:

- Asking
  Like a child who demands that and expects their needs to be granted because they can't do otherwise.

- The give-and-take
  Love becomes a negotiation between merchants—a balance between giving and taking. I am faithful to you if you are faithful to me. I love you so much if you love me so much. I'll be there if you'll be there.

- Donating
  Love is also experienced through donating. The unconditional gift. A parent toward their child.

- Being
  Love is no longer experienced through doing, asking, giving-receiving, or donating in order to enter the sphere of being. So many people do and struggle because they think it's the only way to love or be loved. Here the perspective changes: one is love itself. You become what you want to know.

THE SECRET SCRIPT: EVERY RELATIONSHIP IS A MIRROR OF ONESELF

Many relationships are apparently happy but lack depth and meaning. You meet each other, and share your life and many situations, without ever getting really intimate. The intimacy of a relationship doesn't consist in "telling the problems." Human beings have an unconscious fear of exposing themselves, showing their conflicts and weaknesses to others, and being hurt or abandoned. For this reason, true intimacy consists in both sharing, welcoming, and integrating one's own and others' shadows, and revealing the most authentic and profound part of oneself, without any more fears. Transforming a relationship into a conscious encounter means recognizing in it a fundamental opportunity to get to know oneself. Unlike the superficiality of a common relationship, in the conscious one you go deep inside yourself; the other becomes a tool to identify and resolve one's own shortcomings as well as recognize unexpressed potential. The relationship becomes a conscious choice. Every relationship starts from a polarity: from the perception of someone other than oneself toward which one feels attraction, repulsion, or neutrality. In any case, in a more or less explicit and visible form, there will be a relationship of interdependence and inter-connection just for the simple fact of existing. In the instance of attraction or repulsion, the relationship passes through an integration process: through the other, one has the opportunity to know and feel oneself. The relational one is therefore a process of integration through which one experiences aspects of oneself that must be recognized, understood, integrated, and resolved (even if that integration is not necessarily achieved). We can identify four macro phases:

1. polarity
2. attraction/repulsion
3. relationship
4. integration

Relationships allow us to experience aspects of ourselves that we project outward and that others reflect: in friendship, we find in others common qualities and values with which we identify; in hostility, aspects of ourselves emerge that we reject or deny; and in admiration, we find the seed of potential qualities never expressed or developed. We project desires, hidden expectations, and needs onto the other that are never satisfied, and force us to make a choice: to look elsewhere, in another relationship, or face and try to solve the problem.

Lacking appropriate responsibility, awareness, and inner tools, relationships usually end in disappointment, anger, rejection, and pain. There is a fundamental question that all human beings should ask themselves before delving into a relationship. It is a seemingly simple query yet capable of stimulating great awareness: What does this relationship represent to me? Does it represent security, strength, protection, submission, a need for self-assertion, or recognition? What deep needs does it satisfy? Is it an anesthetic for fear, loneliness, or insecurity? What is its deep purpose? Does it represent the integration of a father you never had, a mother full of kindness, or a son to protect? The essence of the relationship is contained in the honest answer to this simple question: What does it represent to me?

If you have enough patience to get to the bottom of it, repeating the question until you leave behind the superficial and partial answers, an unequivocal clarity will emerge about the deepest meaning of that relationship. Beyond the thousand facets, the daily and medium and long-term goals and objectives, there is a much more intimate and purpose, which represents the true meaning of that encounter. That's the real glue. If it were to fail or be exhausted, even for just one of the two people, the relationship would fail. You need to get naked not only physically and look in the mirror without hiding your real expectations and needs.

The deep purpose may not be the same for the two people; for this reason, it's good to try to discover and clarify one's own reason, and possibly foresee or understand the partner's. Almost everyone has a "secret script" when entering a relationship; there are few people who are totally transparent and with nothing to hide.

These are hidden expectations, frequently denied even to oneself and not declared. It could be the desire for a child, to create a family in a certain way, or conversely, not wanting a family or children at all. The secret script is about our inner movie—what we really expect from that relationship. There will come a time when we will have to declare it. And accept it. There are relationships that go on for years and never get to this point. They postpone, without ever listening to what is written in that script, often trampling on their own desires in the name of the other's needs or for the survival of that bond. One way to start connecting with the secret script is by asking yourself the question, What does this relationship represent to me? The secret script concerns what you really want from that relationship; there are people who are inclined to take or obtain others, instead giving or being recognized, because they badly need support that makes them feel safe. It's not about reducing relational dynamics in simplistic terms but rather about beginning from clarity and inner awareness, in order to guide choices, decisions, behaviors, and sentimental investments.

Will we have the courage to admit to ourselves all the compensatory mechanisms that we have projected onto the other person and love each other anyway? How much do we really care about the other and how much are we centered on ourselves instead? A lot of people can't get out of a long, already-ended relationship for fear of stepping out of the character they're playing. Some relationships merely end when they've done their job. Some couples are formed only to bring a child into the world, and when that task has been fulfilled, the meaning of their journey together is complete and it's time to split up; it would no longer make sense to continue. Yet if there's no awareness and courage, separation will come with much more pain. Other relationships will last a lifetime, however, since that is their purpose. How do we know if a relationship has finished its task and it's time to change? How do we understand when it's appropriate to continue to stay together and insist despite the hurdles?

Let's go back to the starting point: in order to know when and if a relationship has come to an end, you must first clearly know the meaning and purpose for which it began. Was it clear from the beginning, did we discover it along the way, or are we discovering it right now, in the face of a crisis? Being able to understand what that person and the relationship represent allows us to understand if that need has been satisfied, or if there is frustration and a block, and thus decide whether to continue or not. So the first step consists in becoming aware within oneself of what that relationship represents. And then, only later, ponder your choices.

There are more superficial reasons and then superior archetypes that act and guide our decisions. The person we have in front of us and with whom we have chosen to establish a relationship responds to precise profound needs for security, stability, beauty, passion, freedom, strength, protection, sweetness, recognition, and acceptance, but also the parental figure in the aspects that we have never known or a child we never had. The possibilities are endless and respond to personal needs. What is certain is that the person in front of us and the meaning of that relationship are celebrated on this archetypal level too. Ultimately, we relate with ourselves, or rather, with those aspects that the other represents and we need to experience, in order to integrate, know, and possibly accept and love them. This brings clarity and honesty within ourselves, then improving the quality of relationships, and directing them toward greater harmony and balance.

LOVE AND FALLING IN LOVE: WHERE IS THE TRUTH?

It's among the most studied, mysterious, and least understood subjects, at least until recently, when intense research began: How does falling in love really work? More than twenty years ago, biological anthropologist Helen Fisher studied 166 populations around the world and found evidence of so-called romantic love—the kind that leaves you breathless and with butterflies in your stomach—in 147 of them. This suggests that there are profound biological reasons behind this type of behavior. But how does it work? And how does it differ from love? In 2005, Fisher and her research team published a groundbreaking study that included the first functional magnetic resonance imaging (fMRI) of the brains of individuals in love. Her team analyzed twenty-five hundred brain scans of college students who viewed photos of their "sweetheart" and compared them to those taken when the students looked at photos of mere acquaintances. The two photographs had distinct effects on the participants' brains: the sight of the lover's pictures activated the brain areas rich in dopamine receptors, the so-called pleasure neurotransmitter. Two of the brain regions that showed the greatest activity were, respectively, the one related to reward seeking and expectation (the caudate nucleus), and the one related to integrating sensory experiences into social behavior associated with pleasure, focused attention, and motivation to obtain the reward (the ventral tegmental area). This circuit is considered a primitive neural network—that is, from an evolutionary standpoint, belonging to the brain's ancient structure—and it seems that the most primitive areas of the brain are the ones most involved in the process of falling in love. When we fall in love, the chemicals associated with the reward circuit flood our brain, producing a wide range of physical and emotional responses: the heart beats wildly, the body sweats, and the face turns red, and we experience emotions that we label as anxiety, excitement, fear, and exaltation, depending on the context and which ones have in common a significant emotional activation. Levels of cortisol, the stress hormone, skyrocket, preparing the body to cope with "crisis" or "danger." Put simply, our brain doesn't distinguish between the feeling of falling in love and coming face-to-face with a wild beast, and to be safe it activates an instinctive fight-or-flight response.

But this is not all, because the release of dopamine during falling in love, in addition to activating the reward circuit, causes the same feeling of euphoria associated with the use of cocaine or alcohol. A unique study done at the University of California at San Francisco and published in 2012 in *Science* reported that fruit flies that had been rejected by their companion flies drank four times more alcohol than those that had managed to mate—a reaction now shared and socially accepted by the collective imagination of human beings. Just think of the hundreds of films in which the rejected lover finds themselves alone crying at the counter of a bar. The reward system is basically the same; what's different is the way to reach it. Falling in love is certainly one of the most powerful sensations on earth, and according to Fisher, it has three peculiar characteristics: the intense desire toward "the object of falling in love," the motivation that pushes one to do everything to obtain it, and the obsession toward them that leads all thoughts to converge in that direction. Donatella Marazziti, an internationally renowned Italian psychiatrist, added a fourth fundamental factor. In commenting on one of her latest research studies to Candida Morvillo in *Corriere della Sera*, she states that it takes just six milliseconds to fall in love and twelve milliseconds to know it, thereby concluding that falling in love is an involuntary and transitory phenomenon, while love implies an act of the will. Even science seems to confirm that falling in love is a state of altered consciousness, wonderful perhaps, but transitory. Most of us have had the same experience: that moment in which we are convinced—hand on fire—that we could never live without that person, who is the right one for us; we immediately understood they were the one and are certain that the relationship will be "forever."

Then something happens—science would say that the chemical reaction of the brain normalizes—and that phase of acute falling in love suddenly vanishes like a soap bubble: it passes; we realize that it was a transitory thing and that deep down we immediately understood there was something in that person that was not right for us.

But where is the truth? Was that person the right one or not? What if neither question offers a correct answer, and the person is both right and wrong at the same time? When we fall in love, we project our deepest desires onto the other person, such as the desire to be a parent or be loved unconditionally; we see the unrecognized potentials in us expressed in the other, such as the denied needs and hidden expectations. In this way, we transform the other person into a charming prince or fairy-tale princess, precluding ourselves from seeing, accepting, and loving them for what they really are. Then one day the spell is broken, and we realize that all the expectations projected onto the other are partially or completely shattered and disregarded. At that point we find ourselves at a crossroads: to unload all of our projections, preconceptions, and unfulfilled expectations onto the other, or take full responsibility for what happened. It's the "responsibility" to fully and consciously live the process we are going through, leaving room for a new universe where there are no preestablished scripts, definitions, or crystallized roles. In that moment, when we let go of everything we thought we knew, there they are, the infinite possibilities of existence. The wonder of every moment in which we rediscover ourselves as something new in eternal transformation, we flourish again in the enthusiasm of a new vital flow that nourishes us deeply. Then we recognize ourselves in that state of love no longer tied to a name, giving or receiving, or doing, but composed of the same substance that we are made of. Thus with kindness, falling in love gives us a new possibility by opening the door to true love, the conscious one. The choice is ours.

CHOOSING AND DECIDING IN RELATIONSHIPS

During a ten-day meditation retreat, one of the participants confessed to me that she had great resistance, so much so that she was thinking of abandoning the experience. I asked her what was keeping her from leaving. She told me she felt guilty because she was afraid of disappointing me and the friends she had come with. On the one hand, she felt that she would have lost an important opportunity, and perhaps by remaining and overcoming her resistance, crucial changes and openings of consciousness could have happened. On the other hand, she was frustrated because she couldn't silence her mind and this made meditation a hell for her. Seeing her so hesitant, I asked her if she was willing to play a game with me. I took a coin and said that if it came up heads, she would have stay, and if it came up tails, she would have to leave. She accepted. I tossed the coin up, and when it fell into my hand, I covered it with the other. I looked the person in her eyes and slowly revealed the outcome. "It was tails, you have to go away." She impulsively replied, "But I want to stay!" Her choice was already clear inside her, but she wasn't aware of it. Sometimes, people are aware that they have made a certain choice only after making it, as if one realizes that the other was the right person only when the relationship is already over. We should find the humility to listen, the clarity to understand, and the courage to follow what we have heard.

Every time you are faced with a choice, you should remember the Zen saying, "When you are at a crossroads, take it." What makes the difference is not so much what we choose but rather how aware is the part of us that makes the choice. Who makes the choice? A woman who feels lonely? A frustrated man? An abandoned child? A betrayed person? A free and happy being? Which part of you is choosing? Discovering this is an important advantage because it makes clear the real needs and requirements, and also the compensation mechanisms that prompted us to make that choice.

THE MECHANISMS OF CHOICE

It would be helpful to clarify these four aspects:

1. Why is a choice made? Be clear about the reasons.

2. What is being chosen? Be clear about what it is and be able to describe it in a few words.

3. How does one choose? What are the ways through which one makes the choice?

4. Who chooses? What part of us makes that choice? From what perspective do we choose, what needs are we fulfilling, and what, if any, are the compensatory mechanisms?

FACTORS INFLUENCING AND GUIDING CHOICES IN RELATIONSHIPS

Most people "unknowingly" choose to start or continue a relationship, be it a couple, friendship, or professional or recreational one. *Unknowingly* means that one doesn't have awareness or full clarity of the factors that influence that choice, ignoring the deep reasons that motivate it, including all the compensation mechanisms for fears, shortcomings, insecurities, and so on.

The first factor influencing these choices is the instinctual and sensory one. The choice is made on the basis of instincts and impulses. It's dictated by satisfying sensory pleasure.

The second factor is sentimental. The choice recalls, evokes, and is associated with a feeling, such as affection, security, or familiarity. It's the feeling that relationship evokes that drives and influences the decision. Think of the grade given by a professor who is influenced by the sentimental aspect: the student reminds them of a child or loved one, or they simply like them.

The third factor is intellectual: the so-called reasoned choice, in which the "whys," pros and cons, and advantages and disadvantages are carefully evaluated.

The fourth factor is ideological. At the basis of the choices, there are ethics and values. In this case, priority is given to respect and consistency with one's own values. Let's think, for example, of the value of premarital chastity prescribed by various religions, the value of ethics in choosing a job, or the value of honesty when choosing a friendship.

The fifth factor is social. The relational choice is made according to a social purpose or recognition. These choices are not only influenced by social values such as solidarity, service, and a sense of responsibility to the community but also dictated by the need for acknowledgment and affirmation of one's social position. Some examples of this are arranged marriages between people in show business, or choosing partners, jobs, and friends based on their social prestige.

Finally, the sixth factor is related to awareness. Choices become expressions of one's inner awareness. They are made when there is clarity about which part of oneself is making the choice and what that choice really represents. Being aware of who chooses will be more important than what, how, and why one chooses. Each choice will be used as an opportunity to increase one's level of inner awareness.

THE FOUR QUESTIONS

We can conclude that the quality of relationships starts primarily from honesty, clarity, and inner awareness about one's needs, requirements, and objectives. Being honest with yourself is fundamental for healthy relationships. To test one's degree of inner sincerity, it would be enough to honestly answer these four questions.

1. What part of you makes the choice, and what needs and requirements does that choice respond to?
2. What does that choice represent?
3. If you were totally free and happy, what would you choose?
4. If you knew you were going to die soon, would you still make that choice?

These four questions can greatly discourage a lack of self-honesty. The first question clarifies the real needs and requirements that drive us to make that specific choice, and to which level of identity they belong. The second question sheds light on the purpose of the relationship beyond superficial appearances. The third question makes us aware of the compromises and compensation mechanisms: if, despite being totally free and happy, we redo the same choice, this will prove it's authentic. If we go beyond all the logic of convenience, choice is no longer a search for freedom and happiness but instead becomes an expression of them: you choose and decide because you're free and happy, not because you want to be free and happy. The challenge is to answer this question having the courage to honestly listen to what resonates within us. Without judgment. The fourth and last question is even more effective because it puts us in front of an unavoidable event: death. All that is superfluous and useless falls before it. It's a return to the essentials where only what is authentic and what really matters in our lives remains. There's no more time for lies. There's no longer room for excuses and fears. We are willing to make choices that we've never made before because there's no longer time for shame, lies, the unsaid, and tricks. Life should be lived this honestly.

9 THE SCIENCE OF RELATIONSHIPS

IMMACULATA DE VIVO

> I have decided to stick with love. . . . Hate is too great a burden to bear.
> —Martin Luther King Jr., "Where Do We Go from Here?"

Savanna baboons tend to form strong, balanced, and stable bonds with each other, and those bonds can even last a lifetime. Females in particular have the ability to bond closely with other members of the herd, especially with other females, according to ties of kinship or similar age. The resulting relationships have a great impact on their lives because they influence their ability to react to stress, their chances of reproductive success, and their life expectancy. Females who have good relationships within the pack tend to live significantly longer than those who maintain weaker and more unstable relationships. Having good social relationships is an adaptation factor promoting health and longevity. It's true for savanna baboons, lab mice, and humans as well.

The bonds we establish with other people influence our emotions and mental balance, but they go even deeper than we could imagine. They penetrate under our skin, impacting the biological processes that keep us healthy or make us ill. They come to lengthen or shorten our very existence, so great is the force they exert on many basic aspects of our lives.

GOOD RELATIONSHIPS, GOOD HEALTH

The nature and importance of social relationships for an individual's well-being are particularly interesting for medical science and biology, because they not only give us pleasure and mental satisfaction but can also influence overall health. They exert a surprising long-term effect on our body, comparable to other healthy habits such as sleeping well, eating healthily, and not smoking. There is now an extensive scientific literature on the subject, telling us that people who have adequate social support, from family, friends, and community, are happier, enjoy better health, and live longer.

In 2010, the scientific journal *PLOS Medicine* published a report of 148 different studies, for a total of 308,849 subjects observed. The results of this large amount of research show that the lack of strong social ties can increase the risk of premature death by up to 50 percent, regardless of factors such as age, gender, or health status. That's a high percentage, comparable to the mortality risk for a person who smokes fifteen cigarettes a day, and even higher than those who are obese or sedentary. Social isolation therefore has a toxic effect on our body, and is responsible for a greater susceptibility to various types of diseases and illnesses.

Scientists are currently interested in analyzing the biological and behavioral factors involved, such as stress levels. The prolonged presence of hormones like cortisol in the blood can cause damage to the cardiovascular system as well as interfere with gastrointestinal functions, insulin regulation, and immune response. It's a combination of physiological reactions influenced by psychological factors, on which research is still ongoing.

Among the many scholars who have studied this matter, Robert Waldinger undoubtedly stands out as one of the world's leading authorities in the field of psychiatry linked to the study of human relationships. A professor at Harvard University, author of books, and speaker at famous TED conferences, he's also my personal friend, with whom I've conducted studies in collaboration. In his long career as a scholar, Waldinger analyzed various aspects of social life, especially how human relationships can influence an individual's well-being. For many years he directed the Harvard Study of Adult Development (HSAD), a study begun in 1938 and still ongoing, following the emotional and social lives of 724 men, about 60 of which, now over ninety, have been in the project from the start. The participants were divided into two groups: the first, recruited in their sophomore year at Harvard, finished their studies during World War II, when many left for the front; the second group was made up of boys from Boston's poorest neighborhoods chosen because they belonged to particularly disadvantaged families. In these eight decades of research, it's been possible to accumulate an enormous amount of data, also regarding the wives and children of the study subjects. If the outcome of this study were to be summarized in a few words, according to Waldinger the fundamental message would be, "Good social relationships make us happier and keep us healthy."

More specifically, we can draw three key lessons from this research. The first is that social relationships are good for our well-being, while loneliness can kill. Having a dense network of relationships, in the family, with friends, and with the community in which one lives, has a positive impact on various good health indicators and lengthens life span compared to those who live more isolated. Loneliness is not only the lack of relationships but also the inability to enjoy the ones you have, as happens to those who feel alone even in the midst of a crowd or their marriage. The second lesson is that when it comes to relationships, quality is always more significant than quantity. In friendships, for example, it's more important to have a "few but good friends," as popular wisdom says, and scientific data confirms this. In romantic relationships, an excessively conflicted marriage, with too much fighting and the lack of a strong bond to compensate for it, has proved toxic to mental and physical health, sometimes making divorce preferable. The third lesson is that good social relationships, especially couple relationships, improve brain performance even in old age and can protect functions that tend to fade with age. Memory, for instance, remains lucid longer, whereas people who can't count on a partner show early signs of memory decline.

HSAD has provided a lot of information to scientists about the value of relationships along with what can or cannot influence our ability to form good and lasting ones. For example, it was found that the family origins of the study participants didn't have much influence on the quality of their relationships in adulthood: coming from a poor or wealthy family, being successful or going through a lot of turbulence, doesn't seem to count for much. The protective effect of good sociability is maintained over time: if a person is satisfied with their relationships at fifty, they will be in better health at eighty.

Data relating to eighty-year-olds who declared themselves happier in their relationships showed that their state of mind remained positive even on days in which they suffered from physical aches. This comes from a well-known 2010 study, titled "What's Love Got to Do with It?," in which Waldinger examined a sample of forty-seven octogenarian couples; day by day, he observed correlations between perceived well-being and happiness in relation to time spent with the partner or other people. Satisfaction with marital life and time spent with others both positively influenced the sense of well-being and happiness of the observed subjects, who derived satisfaction from both types of relationship. But in case of declining health, only the couple's relationship was able to mitigate the resulting lowering of happiness levels, while time spent with others didn't have the same effect. The marital bond proved to be more powerful in protecting elderly's happiness from fluctuations due to changes in their perceived health, but it also confirms the importance of bonds with other people in maintaining adequate levels of satisfaction.

In another 2015 study, Waldinger specifically examined the effects that stable relationships have on the cognitive and emotional well-being of married people, even in old age. Analyzing various indicators of mental and physical health in eighty-one elderly heterosexual couples over a period of two and a half years, he observed that the greater sense of security linked to the stable relationship determines greater satisfaction with the relationship, lower incidence of depressive symptoms, improved mood, and a lower frequency of conflicts and quarrels—data that were confirmed stable over time until the end of the observation.

As already pointed out by Waldinger, this doesn't necessarily mean that couple relationships are better than other types of human relationships because it depends on the quality factor. Maintaining a network of "healthy," nontoxic, and fulfilling social relationships is a matter of personal choices too. All relationships have ups and downs, but what's important is to maintain a positive consideration of a certain relationship, knowing that the person in question is a reference point you can count on in times of need. For instance, Waldinger highlights how after retirement, many elderly people risk losing friendships because they no longer have a daily relationship with work colleagues. It's crucial in this phase of life to frequent new environments, such as associations or volunteer groups, in order to make new friends. The HSAD showed that the happiest men in retirement were those who had replaced the long-lost ties with colleagues with new, equally close and solid friendships. According to Waldinger, being connected to a social network of healthy relationships is a way to take care of yourself, like exercising or eating well. It's a true prescription for good health.

Speaking of physical health, social relationships have an important impact especially on cardiovascular functions. A study from the Harvard School of Public Health led by Ichiro Kawachi followed a population of 32,624 men between the ages of forty-two and seventy-seven who had no coronary heart disease, stroke, or cancer for four years. Individuals who were more socially isolated—unmarried, with fewer than six friends or relatives, and not members of parishes or other local gathering groups—had a higher risk of death from cardiovascular disease, but also from accidents or suicide, and were more likely to suffer from nonfatal strokes. Having a strong social network was instead associated with a lower risk of stroke and higher survival rate after the onset of coronary heart disease.

This correlation was also observed in a 2014 study at Harvard University in collaboration with various California universities in relation to the survival rate in women with breast cancer. The study involved a sample of 2,835 women from the NHS who had received a diagnosis of stage 1 to 4 breast cancer. Their networks of social relationships were observed, and by crossing data, independent of various potential confounders, it was seen that the most socially isolated women before diagnosis had twice the risk of death from cancer than their more socially integrated counterparts. In particular, the greatest risk concerned women who did not have close relatives, friends, or children—a fact that was found in the analysis carried out after the diagnosis too. Researchers hypothesized that the main reasons are late access to medical care along with a lack of support and closeness from friends, relatives, and adult children.

DOING GOOD TO OTHERS IS DOING GOOD TO YOURSELF

The impact on health of the particular relationship between a sick person and the one who takes care of them has been investigated in various scientific studies. Taking care of someone in a difficult situation, deprived of their autonomy, can have mixed consequences on the caregiver's psychology: on the one hand, it can increase the load of stress and cause greater susceptibility to disease, undermining overall health; on the other hand, it sets in motion mutual exchange mechanisms based on love and emotional gratification that can greatly compensate for stress factors, thereby lowering risk rates. The matter is still under observation, but there are already important results on the beneficial effect that taking care of a loved one can have on one's health. In the United States alone, it's estimated that approximately 21 percent of the adult population takes care of a person over the age of eighteen free of charge. This doesn't include professionals and childcare providers, thus allowing us to limit our research to that specific type of caregiver-patient relationship that is established between two adults out of an emotional bond. A study at the University of Michigan looked at a population of 3,376 married adults over the age of seventy engaged in spousal care. The results showed that those who provided assistance for at least fourteen hours a week had a lower mortality risk. The most likely explanation is that emotional gratification and the affection bond can reduce perceived stress, neutralizing the opposite effect exerted by the difficulties and worries related to the caregiving.

An international study conducted in Berlin and published in 2017 confirmed these results, highlighting the benefits of caregiving in a sample of 516 people aged eighty-five on average. The research showed an improvement in health conditions and longevity in these subjects, devoted to taking care of other people in their own family or within their social network. Among the possible reasons, hypotheses were made that people who are committed to helping others, even in old age, play a useful role in evolution because they guarantee the persistence of ancestral functions linked to being parents or grandparents—protective toward the society to which they belong. This would have led to a particularly honed neuronal network and hormonal system to support prosocial behaviors even outside the close family circle. Helping others becomes a mechanism for aging well and guaranteeing one's participation in social life, obtaining "in exchange" an extended life expectancy.

Even more significant is the beneficial impact of volunteering, which doesn't imply a strong emotional relationship between the subjects involved and is usually aimed at a plurality of individuals. Carrying out free activities to help people who often don't even know each other is an act of pure generosity, with no return other than the satisfaction of having done good for someone else. For years I've been doing a few hours a week of voluntary work in a food pantry for disadvantaged people in my community and had the privilege of experiencing firsthand the bond of profound mutual respect that you establish with the people involved as well as the sense of gratitude that holds together the parties of the exchange. Science investigated the effect of this important emotional experience on general health, with significant results. A California study analyzed 5,630 people involved in volunteering and observed that the most active subjects, like those who served for more than one organization, had a 60 percent lower risk of mortality than those who performed no activity. The result was confirmed by a 2018 study carried out in Hong Kong in which scientists saw that in a population of 1,504 adults, those who volunteered had better health, specifically a lower incidence of depression, better physical well-being, higher levels of satisfaction with one's life, and greater social integration. Another study found a beneficial effect on blood pressure, which was lower in a sample of people over age fifty observed for four years, likely due to the antistress impact of engaging in this type of activity.

Volunteering has also been studied in its correlations with particular psychological conditions, such as those of war veterans diagnosed with PTSD. One study observed a group of 346 veterans returning from the first phase of the US military intervention in Afghanistan in the aftermath of September 11, 2001, terrorist attacks. The study's participants presented with a wide variety of physical and psychological health conditions, with different levels of emotional difficulties, symptoms of PTSD and depression, sense of purpose in life, social isolation, and perceived availability of support from others. Volunteering has been associated with significant improvements in mental and physical health, regardless of different baseline conditions. The beneficial effect was seen in people enduring strong pressure and stress factors too, with considerable emotional and psychological problems. Turning one's attention and efforts to someone else's well-being has eased the suffering and symptoms of mental distress in discharged veterans, providing further evidence of how effective altruism is in improving the health of those who do good for others. The emotional mechanisms involved here imply a sincere altruistic vocation, as highlighted by a 2012 study: the protective effects are more marked in those who volunteer out of a pure spirit of dedication to others and not to obtain personal gratification.

According to Aristotle, one of the pillars of existence is "serving others and doing good." Such a principle, according to scientific research, could also be fundamental for maintaining good health and living longer.

THE TEN BENEFITS OF VOLUNTEERING

1. It creates a sense of community

Volunteering strengthens bonds within a community, whether it's a village, small town, or neighborhood in a large city. You make important connections with the recipients of your assistance and other volunteers.

2. It's an antidote to loneliness

Studies say that nearly 45 percent of people in the United States and United Kingdom report feeling lonely, and one in ten people say they have no close friends. Volunteering allows you to get out of loneliness and get in touch with people.

3. It creates bonds and friendships

Good sociability helps improve mental and physical health, with surprising long-term effects. Brain functions are strengthened, as is the immune system, while stress, anxiety, and depression drop. Sharing a charitable activity is the best way to create new friendships and long-lasting relationships, based on common values and an experience that unites.

4. It promotes emotional stability

Volunteering has proven effective in alleviating the symptoms of various emotional illnesses, as it lowers stress and offers a strong sense of purpose that mitigates emotional instability.

5. It promotes longevity

Volunteering can lengthen your life as a result of the combination of many positive effects that improve mental health, heart health, and overall happiness. And it's a benefit that can be enjoyed at all ages.

6. It reduces the risk of Alzheimer's disease

Doing good for others can significantly reduce the risk of developing dementia, of which Alzheimer's disease is a variant. The brain elasticity is preserved, helping to maintain cognitive functions, especially memory.

7. It promotes good aging

Older people are the ones who benefit most from the positive effects of volunteering. Heart health improves, thanks to the antistress action and physical movement associated with the activity performed.

8. It improves school and university experience

For young people attending school or a university, engaging in volunteering means improving their social skills, learning practical activities, expanding their mind compared to others, facing new challenges, and accumulating experiences. All of these are useful elements that positively influence one's study performance.

9. It improves your professional prospects

Many companies and public authorities hold a high opinion of a candidate's volunteering experience because it highlights their skills and interests along with demonstrating versatility and adaptability.

10. It's fun

When we volunteer, we make new friends, enjoy the pleasure of doing good for someone else, spend quality time with people who share our values, and build pleasant experiences and memories that will last.

10 NUTRITION: HEALTH COMES WITH EATING

IMMACULATA DE VIVO

I don't think any day is worth living without thinking about what you're going to eat next at all times.
—Nora Ephron, "My Day on a Plate"

In the 2007 animated film *Ratatouille*, the protagonist, a mouse-chef named Rémy, prepares a dish for Anton Ego, the most terrible and feared food critic in all of France. Tension is skyrocketing because the restaurant risks seeing its good reputation ruined if Ego gives a bad review. But the moment the critic takes his first bite, his mind is catapulted back in time to the carefree days of his childhood. Times when happiness had the taste of the simple and delicious dishes that his mother cooked, and that turned even the saddest days into memorable ones. It is an experience that has happened to everyone in their life—the time when years later, we have unexpectedly rediscovered a forgotten taste of our happiest memories, reliving for a moment the emotions of that distant past.

This is the power of food to give joy and comfort as well as recollect memories that make us feel good about ourselves and with others. Food is the protagonist of happy moments; it gathers families and strangers, and builds bridges among different cultures. In the right doses and with the right quality, food makes us feel good in many different ways, in body and mind, and has the ability to make us live longer and healthier lives.

Science has long demonstrated the crucial importance of nutrition for our health and well-being—a concept that we must always keep in mind. What we eat literally represents what we are made of, the raw material that makes up our cells and tissues, the vital energy that permeates us as well as shapes our thoughts and emotions. If this raw material is of good quality, so is the biochemical environment that derives from it, therefore ultimately our state of inner well-being.

Food is a vast and complex topic. It gets almost daily coverage by newspapers, TV, and radio shows, and is the subject of ever more numerous and sometimes contradictory books and debates. Healthy discussions and differences in opinions and points of view can greatly propel research and progress, but for the lay public it can sometimes become confusing to be bombarded by conflicting statements. Frequently, we read shocking headlines about new studies contradicting established "truths" and known common beliefs. The more shocking the news, the more attention it gets in the media, especially when scientific truth is expressed in black and white rather than in its true, nuanced form. Scientists tend to be cautious, but this gets lost in the media frenzy. The result is that many people, even the most sincerely interested in the subject, end up believing less and less in scientific studies. They are convinced that science is unreliable and scholars mostly cannot be trusted. We can still grasp a minimum amount of sound knowledge, enough to be convincing, even in the avalanche of new studies mixed with sensational headlines. We scientists are trained to be prudent and not to overinterpret findings, yet in the face of numerous and consistent studies, we can make well-informed and scientifically sound choices.

TELOMERES AND THE MEDITERRANEAN DIET, A LOVE STORY

The impact of nutrition on health has always attracted the interest of various medical disciplines, and in recent years we also enlisted telomeres as among the great protagonists of this research. Because telomeres function as a biomarker of aging, it makes them particularly interesting in order to understand how our eating habits influence our health. To date we have collected a significant amount of data to formulate conclusions and identify the general rules of good nutrition from a molecular as well as genetic point of view.

Let's start from the basic concept that the main enemies of telomeres are oxidative stress and inflammation. These two biochemical conditions accelerate telomere attrition, causing premature aging of the cells and increasing the risk of developing chronic disease. Anything that fights oxidative stress and inflammation is thus good for protecting telomeres. Moreover, such is the case with meditation, optimism, kindness, and forgiveness, which we can define as food for the soul. And this is the case with nutrition—that is, the foods that we introduce into our body to provide it with replacement substances and energy.

The most important nutrients for protecting telomeres contain antioxidants and anti-inflammatory molecules. These are abundant in plant foods: fruit, vegetables, cereals, legumes, dried fruit, and seeds. Every time we eat plants, we supply our body with a large variety of substances that act as a shield against the damage caused by adverse biochemical reactions.

Let's imagine these nutrients as a jet of water that extinguishes many small fires that are consuming our telomeres. Oxidative stress and inflammation impact telomeres in the same way fire devours the ends of a log, slowly but steadily reducing some sections of our telomeres to ashes, or more than their typical shortening rate.

Fiber, present only in plant foods, plays an important role in protecting telomeres too. A 2018 study conducted through the National Health and Nutrition Examination Survey analyzed DNA samples from 5,674 US adults, measuring their daily fiber intake. In general, people in the United States have a low intake of fiber—lower than the US Department of Health's recommendations. The study found that for every gram of fiber added per thousand kilocalories, telomeres were 8.3 base pairs longer. Since one year of normal cellular aging in the reference sample corresponds to a shortening of 15.5 base pairs, it has been calculated that an increase of ten grams of fiber per thousand kilocalories corresponds to 5.4 years less of cellular aging.

The protective action of fiber in any case corresponds to about 4.3 years less of cellular aging, even accounting for factors independently affecting telomere length, such as smoking, BMI, and physical activity. Fiber has therefore shown a strong correlation with longer telomeres, confirming, including from a molecular point of view, the main rule of healthy eating: base one's diet on plant foods as much as possible.

The traditional Mediterranean diet, abundant in vegetables, fruit, legumes, and cereals, encapsulates the main foundations of a healthy diet. Due to its exceptional uniqueness, it has been the focus of vast scientific research and has proved to be a formidable telomere protection factor. Recognized by the UN Educational, Scientific, and Cultural Organization (UNESCO) in 2010 as an intangible heritage of humanity, this diet is characterized by a high consumption of plant foods, moderate intake of olive oil and red wine, and marginal intake of meat, dairy products, and saturated fats. This combination, a result of a food culture that has developed over millennia in some Mediterranean countries such as Italy, Greece, and Spain, has proved to be scientifically valid. It has been shown to counter oxidative stress and inflammation, thereby reducing telomere wear and tear, and representing an effective way to promote longevity. It is no coincidence that two of the five Blue Zones in the world, defined by a unique concentration of centenarians, are located in the Mediterranean basin: Sardinia in Italy and Ikaria in Greece.

The first scientific insights into the benefits of this diet date back to the end of the 1950s, when US physiologist Ancel Keys started the Seven Countries Study on the incidence of cardiovascular disease in different countries of the world. Still ongoing, the Seven Countries Study was the first major epidemiological research to link this type of disease with eating habits, paving the way for an infinite number of studies that have led us today to have vast knowledge on the link between the foods we eat and our health.

From Keys's observations, updated by more recent studies, it has been shown that cardiovascular diseases mainly affect the populations of North America and northern Europe, while countries in Mediterranean Europe and Japan have a much lower incidence. Research has made it possible to link the onset of these diseases with a high consumption of saturated fat and high blood cholesterol levels. The Mediterranean model, based on a massive consumption of vegetables and low intake of animal products, has been associated with a lower incidence of cardiovascular disease and lower mortality rate.

Following this food pattern, then, means increasing the chances of living longer and healthier. The most recent studies have tried to understand the underlying physiological mechanisms, knowing that oxidative stress and inflammation are attenuated by consuming foods typical of the Mediterranean diet. Attention has gone beyond cardiovascular disease to include other chronic diseases typical of aging that can be associated with abnormal telomere attrition. These diseases include diabetes, cancer, Alzheimer's, and dementia. Can the Mediterranean diet lower the probability of developing these diseases, slow down cellular aging, and promote longevity? Research suggests so, and the number of studies on the subject is increasing year by year.

One of the most significant results comes once again from the NHS, which analyzed the telomeres and eating habits of 4,676 healthy US women. It turned out that the greater the adherence to the Mediterranean diet, the longer the telomeres. And it's important to underscore that it's not single nutrients that have proved to be protective but rather the diet as a whole.

A well-known example is the study conducted in 2000 on the properties of a vegetable pita sandwich, a typical Greek dish. The seven wild herbs most used in Greek cuisine and in particular in this recipe have a high concentration of antioxidants, especially flavonoids. These herbs are fennel, chives, wild thistle, Mediterranean hartwort, poppy, common dock, and Queen Anne's lace. The traditional use of these herbs for the preparation of a vegetable pita sandwich involves their combination in a single course, which has been shown to have exceptional nutritional properties. A hundred-gram portion of a vegetable pita sandwich, for instance, contains a quantity of quercetin twelve times greater than a hundred-milliliter glass of red wine. This type of synergy is also present in the typical Santorini broad bean: when mixed with capers, flavonoids are increased sixfold.

The interaction among the various nutrients, even different components of the same food, tends to be more effective synergistically when intact. This is true in the case of whole grains, traditionally used in the Mediterranean diet, but over time they have become less used than refined equivalents—that is, deprived of the outer layers of the grain. The refining process reduces the nutritional value and weakens its protective effects on health, as in the case of heart health, due to the loss of fiber, folate, and vitamin E. Studies conducted in the 1990s have highlighted that the risk of heart disease can be reduced by 20 percent if you consume at least one portion of whole grains per day. Whole grains like popcorn, brown rice, whole grain breakfast cereals, and bran offer the best protective effect.

THE NUTRIENTS OF THE MEDITERRANEAN DIET

The traditional Mediterranean diet is rich in monounsaturated fatty acids, antioxidants, carotenoids, vitamin C, tocopherols (vitamin E), polyphenols (especially flavonoids), anthocyanins, other vitamins and minerals, and dietary fiber.

(continued)

> The overall fat content is about 40 percent in Greece and 30 percent in Italy. Cereals are essentially whole, or in the form of leavened bread or pasta cooked al dente, such as to lower the glycemic index and glycemic load. In addition to the abundance of phytochemicals, with anti-inflammatory effects, plant-based, whole, and as little processed foods as possible also provide prebiotic fibers, which promote intestinal health.

To measure adherence to the Mediterranean diet, Antonia Trichopoulou, a famous scholar at the University of Athens, Greece, has developed the Med Score, a scale that divides foods into "positive" or "negative" categories according to the impact they have on health, and gives each food one point if positive and zero points if negative. The result is placed on a scale ranging from zero to nine, and gives us a useful benchmark for quantifying the adherence to a Mediterranean diet.

We calculated that the difference in telomere length for each point on the Med Score corresponds to an average of 1.5 years of aging. It doesn't mean that the person will actually live that much longer but instead that their cellular aging is slowed down in a proportion corresponding to 1.5 years.

It's an astonishing result because for the first time it allowed us to translate the beneficial effect of this diet on health into numbers. The Mediterranean diet "writes" our food history in our DNA, leaving a mark of well-being and longevity in our cells that comes from tradition, and that we must protect at all costs. If there were doubts about the importance of nutrition for our health, these studies certainly help to dispel them.

TRADITIONAL MEDITERRANEAN DIET IN GREECE

An increase of just two points on the Med Score scale is associated with:

- 25 percent less mortality for all causes
- 33 percent reduction in coronary heart disease mortality
- 24 percent reduction in cancer mortality

The diet consists of vegetables, legumes, fresh fruit, dried fruit, cereals, and fish.

Our DNA is responsive to healthy eating. Preserving DNA integrity by making wise food choices is our responsibility. Healthy eating must be integrated as a lifestyle, enter our culture of well-being, and not be considered a temporary and ephemeral idea of a "slimming diet" or "detox week," which by definition implies that we will eventually return to "toxic" habits. The benefits of a healthy diet are a long-term solution, not only for protecting telomeres from excessive wear, but for normalizing blood values, stabilizing energy levels, and regulating metabolic function.

EFFECTS OF THE MEDITERRANEAN DIET: RESULTS OF CLINICAL STUDIES

The Mediterranean diet aids in the prevention of cardiovascular diseases such as heart attack, stroke, and death from cardiovascular events through the following:

- Diabetes prevention
- Lowering hypertension
- Improved cognition, slowing age-related decline
- Lower incidence of depression
- Long-term efficacy in weight loss in obese people
- Improved sleep
- Improvement of sexual functions in all sexes

Coffee is another protagonist of the Mediterranean lifestyle. In 2016 at Harvard Medical School, we completed a study that hypothesized that coffee, rich in antioxidants, may also be involved in protecting telomeres from accelerated shortening. We know that this traditional drink is associated with good health and a lower risk of mortality, but its effects on telomeres had not yet been studied. Drawing from that inexhaustible and precious mine of data that is the NHS, the pride of Harvard Medical School, we analyzed 4,780 women and found that higher total coffee consumption, albeit taken in different forms, is significantly associated with longer telomeres. This is a pioneering study, but given the large sample size, the results are meaningful. Additional studies should certainly be done in the future. A lifestyle based on a high adherence to the Mediterranean diet and daily consumption of coffee seems to benefit our telomeres, promoting health and longevity.

VITAMIN D AND TELOMERES

Vitamin D, a fat-soluble vitamin, is a steroid hormone that governs the expression of over a thousand genes, thus controlling more than a thousand different physiological processes. Among its properties, it helps the body absorb calcium, playing a role in bone health and muscle strength. It has a protective effect on the heart, fights inflammation, and aids the immune system.

The primary source of vitamin D is sunlight, or UVB radiation emitted by the sun. Scientists have observed that even fifteen minutes of daily exposure to the sun's rays may be sufficient to meet the daily requirement for this vitamin, but it can depend on various factors, including the latitude where you live and how much time is spent outdoors. To a small extent, vitamin D can be obtained from foods such as fish, which is the best food source, but it's not uncommon for deficiencies to occur.

Vitamin D insufficiency can be problematic, particularly with advancing age, as aging decreases the ability of human skin to produce it. This deficiency is linked to metabolic diseases such as type 2 diabetes and obesity, depression, and cognitive decline as well as an increased risk of catching the flu or developing autoimmune diseases like multiple sclerosis, type 1 diabetes, rheumatoid arthritis, and autoimmune thyroid disease.

Vitamin D is also known for its ability to reduce systemic inflammation and cell proliferation. Since inflammation from tissue damage and increased cell proliferation accelerate telomere shortening, vitamin D can reduce telomere attrition through anti-inflammatory and antiproliferative mechanisms.

THE "FOUR PILLARS" OF THE MEDITERRANEAN DIET

Trichopoulou is today considered the "mother of the Mediterranean diet" because she was one of the first scientists to promote studies on it and become its ambassador in the world. Decades of research on the subject have allowed her to outline its four fundamental pillars: it is healthy and nutritious, it's sustainable, it has a high sociocultural value, and it has a positive impact on local economies.

With regard to the exceptional nutritional properties, the current scientific literature is rich in studies that confirm how this diet can prevent the early onset of various diseases and more generally maintain an optimal state of health. The diet helps maintain an optimal weight and reduced waist circumference, lowers the incidence of metabolic syndrome and type 2 diabetes, improves the level of cellular aging, and delays the processes of cognitive decline linked to Alzheimer's or dementia. But it's also a sustainable diet because it has a lower environmental impact than other diets. Being largely based on fruit and vegetables along with a low intake of animal products, it has a reduced impact on water, soil, and energy consumption as well as greenhouse gas emissions, where other typically Western dietary patterns contribute instead to increase each of these indicators. It fights waste because it implements a culture of reusing leftovers in a virtuous process of optimizing resources. It protects biodiversity too, making the Mediterranean basin an area with an exceptional variety of endemic species, many of which are now threatened in their habitats. It promotes the social and cultural value of food in accordance with the very etymology of the word *diet*, which comes from the ancient Greek *díaita*, or "lifestyle." According to UNESCO, the Mediterranean diet is "a set of skills, knowledge, rituals, symbols and traditions" that go from the landscape to the table. "Eating together is the foundation of the cultural identity and continuity of communities throughout the Mediterranean basin. It is a moment of social exchange and communication, of affirmation and renewal of family, group or community identity. The Mediterranean diet emphasizes the values of hospitality, good neighborliness, intercultural dialogue and creativity and a lifestyle guided by respect for diversity." Food therefore becomes a vehicle of integration, a tool for expressing acceptance and openness toward the world, and a factor of cultural growth through conviviality. Finally, it's a diet that favors positive effects on the economy of the producing countries since it enhances the local origin of foods and respects their specificity, favoring the conservation and development of traditional products, in a harmonious balance between people and their territory.

In more recent years, studies have identified the communities of the Mediterranean area where the traditional diet is still alive, transmitted, protected, and celebrated by those groups of individuals who recognize it as part of their intangible cultural heritage: Pollica in Italy—in representation of all Cilento—Koroni in Greece, Soria in Spain, Chefchaouen in Morocco, the village of Agros in Cyprus, the municipality of Tavira in Portugal, and the islands of Brac and Hvar in Croatia.

OBESITY, A THREAT TO OUR HEALTH AND THE PLANET

The invaluable legacy of the Mediterranean diet, recognized globally as a common heritage of values and well-being, seems to be under threat precisely in the countries where it was born. Recent studies highlight how in southern Europe, the Mediterranean food model is less and less practiced, especially by younger generations, influenced by new types of food from other cultures. The spread of fast and junk food high in saturated fat as well as snacks made of refined sugars and artificial preservatives has significantly increased the incidence of obesity in children. According to the 2018 report of the World Health Organization, Italy, Spain, Greece, Cyprus, and San Marino are the European countries with the highest incidence of childhood obesity, which affects 18–21 percent of children and young people. That translates into one in five minors—a surprising and troublesome figure.

We consider the United States as the homeland of bad food and obesity, but data tell us something different: younger generations in the Mediterranean area are distancing themselves from the traditional food model to which they belong, while the rest of the world, primarily North America, has begun to adopt and spread it. Trichopoulou speaks of a "food transition" that is affecting southern and eastern Europe, characterized by weights above the ideal, obesity, and chronic disease related to bad nutrition. It's the erosion of the legacy of the Mediterranean diet, causing damage not only to people's health but also the social, cultural, economic, and environmental fabric.

Approximately 30 percent of obesity can be attributed to genetics, leaving 70 percent as a result of environmental factors, and therefore by definition modifiable and preventable. There are a number of good reasons to avoid obesity or try to get back to a healthy weight if you already are obese.

Obesity, defined by a BMI greater than 30 kg/m^2, has been associated with the early onset of numerous illnesses and diseases. According to data provided by the Italian Ministry of Health, it is a risk factor for chronic diseases such as type 2 diabetes mellitus (44 percent of cases), cardiovascular diseases (23 percent of ischemic heart disease), and different types of tumors (up to 41 percent), and is the cause of death for at least 2.8 million people worldwide every year. The World Health Organization calculates that the incidence of obesity globally has doubled since 1980, affecting

11 percent of the adult population (over 200 million men and 300 million women) and an increasing number of children (over 40 million under five years of age).

Obesity is an important factor in accelerating telomere attrition. A 2018 article published in the *American Journal of Clinical Nutrition* compared the results of eighty-seven different studies for a total of 146,114 individuals observed. Obesity is notoriously linked to the onset of chronic diseases typical of aging even in younger subjects, in part because it increases oxidative stress and inflammation. These studies observed the relationship between BMI and telomere length over the life course of subjects divided by gender, ethnicity, and age (three groups for those aged eighteen to sixty, sixty-one to seventy-five, and plus seventy-five). As the BMI increased, telomere length decreased, especially in the younger and white population.

Putting on extra pounds beyond one's ideal weight is known to have negative consequences on one's health, even if it doesn't lead to obesity. In the early 1990s, an interesting series of studies on women undergoing weight cycling, also known as the yo-yo effect, revealed that the risk of hypertension increased when they regained weight. For every 10 pounds (about 4.5 kilograms) of weight gained, the risk of hypertension increased by 20 percent.

Obesity, like malnutrition, falls under the umbrella of bad nutrition, now considered the leading cause of illness in the world. These two phenomena are closely linked to climate change due to the impact that producing food for a growing population has on natural resources and the balance of ecosystems.

Teaching the merits of good nutrition is essential to promote good individual health, and is fundamental to safeguard the environment and ensure sustainable production too. Favoring plant-based diets, rich in whole and unprocessed foods, is a choice that has proved to be scientifically valid because it protects the body against the most widespread chronic diseases and premature aging. But it also has a positive impact on the health of the planet because producing plants consumes much less water and soil than those required for animal products, and as such, releases fewer greenhouse gases. The interconnection between us and our habitat is tighter than ever, and even a small change in our food choices can have a huge impact on the future of the earth.

11   ON THE WAY TO WELL-BEING

IMMACULATA DE VIVO

We don't stop exercising because we grow old. We grow old because we stop exercising.

—popular saying

Let's forget for a moment the present world, with its comforts, technological discoveries that spare us any effort, and constant commitment to making us perform as many actions as possible while remaining perfectly still. Let's think back to that savanna of a hundred thousand years ago where we met the hungry lioness and try to broaden our gaze to the entire habitat that surrounds us. Any of our activities, whether related to survival or not, are based on the ability to move and the efficient use of our muscles. Gathering wild fruits, growing crops, hunting, finding water sources, running away from danger, fighting an enemy, and exploring a new territory—it all depends on our body, which a million years of natural selection have made perfectly suited to face the environment in which it lives.

Over millennia and progress, this agile and nimble machine, capable of running, climbing, jumping, and crouching, has begun to progressively reduce its need to move, while making living more comfortable. Humans have invented animal-powered means of transport and plowing, designed systems for lifting weights effortlessly, and found a way to exploit the energy of natural forces for various purposes. With the Industrial Revolution, the race toward a more comfortable life led us in just over two centuries to an incredibly sedentary existence. Most of us are able to accomplish an entire day of activities without ever getting up from the chair.

But we are not designed for this. Natural selection provided us with a body that always wants to be on the move, and asks us to walk, run, and jump. A few centuries of comfortable living are not enough to change our evolutionary need for physical activity. The body, forced into an unnatural sedentary lifestyle, reacts by becoming weak, aging prematurely, and falling ill. The only way to counter this unnatural state is movement; it restores the functionality of muscles and organs for which they have been optimized.

SPORT, HEALTH, AND TELOMERES

The benefits of physical activity on the body have been known forever and have been part of our culture since Greco-Roman times, when physical education was a fundamental component of young people's education. Current research allows us to take this knowledge literally into our cells and observe the mechanisms by which movement translates into well-being.

An important Anglo-American study in 2008 analyzed the telomeres of 2,400 volunteers engaged for twelve months in a leisure exercise routine of varying intensity. At the end of the observation, a higher level of activity corresponded to significantly longer telomeres compared to the less active subjects. In 2012, at Harvard Medical School laboratories, we analyzed the DNA of 7,813 women from the NHS, looking for associations between physical activity and telomere length. The results of this study, adjusted for age, BMI, and other possible confounders, showed that women who engaged in moderate or intensive physical activity had longer telomeres than women who engaged in occasional physical activity or were totally sedentary. This difference is translated into approximately 4.4 years of aging, as measured by telomere loss. This is comparable to the difference in aging observed between smokers and nonsmokers (4.6 years). Additionally, longer telomeres were found in women who performed moderate to vigorous physical activity approximately two to four hours per week, matching the current US guidelines. This antiaging effect probably comes from the ability of physical movement to lower psychological stress as well as reduce oxidative stress and inflammation, and therefore buffer accelerated telomere attrition.

This beneficial action was observed not only in healthy subjects with an average weight but also in obese ones. In 2019, a study at the University of Taiwan in collaboration with Harvard Medical School analyzed the effect of physical activity on a sample of 18,424 individuals aged between thirty and seventy years. The participants' obesity levels were measured by five indicators: BMI, body fat percentage, waist circumference, hip circumference, and waist-to-hip ratio. The aim was to understand whether a genetic predisposition to obesity can be mitigated by physical activity and which exercises in particular are more effective. Overall, regular physical activity has been shown to mitigate genetic predispositions toward the first four indicators of obesity. Jogging has been shown to be particularly effective, followed by mountain climbing and activities such as walking, dancing, and long yoga sessions. The conclusion of the study is that regular physical activity can have a greater beneficial impact on those who are genetically predisposed to obesity.

Other scientific evidence comes from the NHS, which over the years has analyzed various aspects of physical activity related to women's health status, such as the link between the intensity of physical exercise and prevention of coronary heart disease. According to statistics, walking is the most performed activity by US women, while a smaller percentage is dedicated to more intensive types of physical activity. The study showed that the risk of heart disease is reduced both among those who walk briskly for at least three hours a week and those who perform intensive activity for at least an hour and a half a week. In both groups, there was a risk reduction of 30–40 percent compared to sedentary women.

A similar type of study was conducted investigating the risk of developing type 2 diabetes and demonstrated that it's not necessary to perform particularly intense exercise to obtain this protective effect. Even a slight increase in one's level of physical activity, such as simply walking versus being sedentary, is enough to lower the risk even in women who have never exercised. The impact is comparable to aerobics or running—that is, about 40 percent less likely to develop type 2 diabetes. This is great news for people who are too lazy to embark on strenuous training sessions, yet motivated enough to improve their health by committing to walk for at least three hours a week. It seems like a reasonable trade-off to significantly reduce the risk of serious chronic disease.

Movement can impact cancer too. Breast cancer was chosen because it's linked to estrogen, a hormone that is reduced by physical exercise. Results found that being active impacts the risk of developing breast cancer in postmenopausal women more so than premenopausal women; even one hour of physical activity a day has been shown to decrease the risk of the disease by 20 percent compared to sedentary postmenopausal women.

The beneficial effect has also been found in older populations, both in healthy people and subjects suffering from some form of frailty. A review published in 2016 by a team of British researchers noted that the prevalence of a sedentary lifestyle among the elderly is strongly linked to the early onset of disease and frailty. In particular, it has been shown that regular physical activity, from simple low-intensity walking to more vigorous exercises and resistance training, can significantly lower the risk of cardiovascular disease, metabolic illnesses, obesity, falling down, cognitive impairment, osteoporosis, and muscle weakness in older adults with various health conditions.

THE MOST EFFECTIVE WORKOUTS

What are the most effective workouts in terms of health benefits? The topic is complex, and still under study to get a deeper understanding and more precise information on what activities are best. According to Harvard professor of medicine I-Min Lee in a *Harvard Health Publishing* article, there are activities suitable for everyone, regardless of age and fitness level, that have been shown to be particularly beneficial to health. These activities help keep weight under control, improve balance and flexibility, strengthen bones, protect joints, and even improve memory performance.

The first of these is swimming, considered the perfect sport. Floating allows you to relieve the load on the joints, especially if they are sore, and promotes a fluid movement. It is aerobic training, so it improves lung capacity and cardiovascular system, and has a particularly positive impact on mood, helping to strengthen mental well-being. Swimming may also include water aerobics, which are useful for burning calories and toning muscles.

The second type of training suggested by Lee is tai chi, an Eastern discipline that is not yet widespread in the West but slowly gaining momentum. This type of activity emphasizes mind and body connection, resulting in an effective state of concentration. Tai chi combines action with relaxation and is referred to as "meditation in movement." It's a series of controlled and graceful movements, with a smooth and gradual transition from one position to another. Speed and intensity change according to the level of expertise, but there are courses suitable for all ages, beginners or veterans. Tai chi offers great benefits especially to elderly by improving balance, thereby reducing the risk of falls or walking with difficulty.

The third activity proven effective for health is strength training, performed by lifting weights and using fitness machines. The idea that weight lifting is only for bodybuilding enthusiasts who want to increase muscle mass is a false belief that needs to be dispelled. A medium-intensity workout, not aimed at hypertrophy, has the advantage of increasing strength and muscle tone without transforming our body. Muscles require large amounts of energy and thus burn more calories even at rest. This type of training helps alleviate age-related diseases too. A study published in the *Archives of Internal Medicine* has shown, for example, that two weekly weight lifting workouts for six months are sufficient to significantly improve memory functions in women with cognitive decline problems. Strength training has also been shown to be effective in relieving knee pain caused by osteoarthritis, improving balance and preventing falls, increasing bone density, thereby reducing the risk of fractures, and improving the quality of sleep.

The fourth workout is by far the simplest to put into practice, but it's also one of the most effective: walking. Scientific studies have shown that walking even just three hours a week has an impact on health comparable to shorter yet more intense workouts. Lee suggests that sedentary people start with just ten- to fifteen-minute daily walks. You can gradually increase, ideally reaching a workout of about thirty to sixty minutes several times a week. Five minutes of slow walking at the beginning can act as an effective warm-up, while five minutes of stretching at the end helps avoid cramps. Scientific research has shown that this type of training helps keep weight under control, lowers "bad" cholesterol (LDL) and increases "good" cholesterol (HDL), strengthens bones, lowers blood pressure, improves mood, and decreases the risk of various chronic diseases.

For sedentary people, an important concept to keep in mind when starting a new physical activity program is that there is no one-size-fits-all recipe. One needs to enjoy the activity to be consistent and reap the benefits. In general, thirty minutes a day of any form of aerobic exercise, combined with two days of strength training, can go a long way toward improving our health.

FOR THOSE WHO DON'T HAVE TIME TO TRAIN

Research has shown that even nonsports activities can offer important benefits. Here are some examples:

- Going up and down the stairs twice instead of just once
- Playing hide-and-seek with the grandchildren
- Parking the car in the farthest place from shopping places
- Walking in place, leg raises, or light weight lifting during TV commercial breaks
- Vacuuming and dusting the house
- Standing and taking a few steps while talking on the phone
- Cleaning and tidying up the garden

TRAINING AND RESTING, THE LIFE-EXTENDING ROUTINE

There's no shortage of scientific evidence to support the effectiveness of physical activity in maintaining good health and promoting longevity. But it's just as important to rest so that the benefits of training are "fixed" into our body. In recent years, in order to understand the ideal number of hours of rest and consequences of sleep deprivation, the number of studies dedicated to sleep and its mechanisms has increased.

What are the main functions of sleep? We know that sleep is fundamental for survival and impacts all systems of our body—our brain, heart, lungs, metabolism, immune system, mood, and resistance to disease. It helps eliminate toxins from the brain accumulated during waking hours. It decreases blood and heart pressure, because the body doesn't need to pump all the blood and the entire circulatory system can recover its strength. It's during sleep that memories are fixed, the information acquired during the day is reworked and selected, and new connections are established that can lead to ideas, discoveries, and even important decisions. It affects the immune system, strengthening the defenses and making them more effective to fight infections. In an interview with Cristina Marrone for *Corriere della Sera,* neurologist Luigi Ferini Strambi, head of the Sleep Medicine Center of the San Raffaele Hospital in Milan, Italy, discusses his study of the impact of sleep on subjects undergoing vaccinations. The subjects who had partial sleep deprivation the day before the vaccination developed much fewer antibodies after three to four weeks than those who slept well in the days around the vaccination.

A lack of sleep has serious consequences on our body and health. Even a night of insufficient sleep can leave us tired and unmotivated, unable to focus on work, exercise, or self-care. Over time, sleep deprivation can increase the risk of developing chronic diseases such as obesity, type 2 diabetes, high blood pressure, cardiovascular disease, cognitive decline, and Alzheimer's, and promote the onset of mental health problems such as depression and anxiety. Vulnerability to cold viruses is also increased. Some brain functions, such as memory and reflexes, are compromised and can likely result in suddenly falling asleep, which is especially dangerous if you're driving. Mood is greatly affected, making us irritable, impatient, and unfocused. Chronic sleep deprivation can cause weight gain by altering the levels of hormones that regulate appetite and affecting the way our body metabolizes nutrients. High blood pressure, high levels of stress hormones, and an irregular heartbeat are linked to a lack of sleep too.

A large proportion of the population is impacted by poor rest, with enormous consequences for public health and safety. In the United States alone, health authorities estimate that over one in three people suffer from sleep deprivation (2016 data). Based on the most reliable scientific evidence, governments recommend at least seven hours of rest for adults in order to guarantee the restoration of all body functions.

Sleep deprivation has also been shown to impact cellular aging and telomere shortening. A 2019 study analyzed the sleeping habits of 482 healthy subjects. Insufficient sleep has been correlated with accelerated telomere shortening, supporting the role of regular and sufficient rest on cellular aging.

My lab at Harvard Medical School studied the correlation between sleep duration and telomere length in a unique population: centenarians. The University of Palermo in Sicily identified an area with a significantly high concentration of centenarians. We examined 143 healthy individuals of different ages (seventy to a hundred and beyond). Telomere analysis revealed, especially in the female population, that ninety-year-olds had telomeres as long as seventy-year-olds. Centenarians who slept an average of eight hours or more a night had longer telomeres than those who slept less than eight hours.

Of course, these data cannot be generalized given the small sample size, yet the findings are intriguing and deserving of follow-up. This was an important discovery, allowing us to investigate multiple correlations. The extraordinary longevity of this "unofficial Blue Zone" provides additional support to an interaction between genetic and environmental factors.

12 THE POWER OF THE MIND

DANIEL LUMERA

The strongest principle of growth lies in human choice.
—George Eliot, *Daniel Deronda*

Hatred doesn't end hatred; it never happens. Hatred ceases through love; this is the "eternal law" that the Buddha taught twenty-five hundred years ago. Over the millennia, there have always been some leaders, saints, prophets, or mystics who reminded humankind that when hatred becomes commonplace, the only true revolution starts with kindness. Today science is doing the same.

Our mind can influence gene expression, modifying aging and inflammatory processes, cognitive abilities, memory, mood, anxiety, and depression. Science and the philosophy of ancient millenary traditions say that genes are not our destiny. We are architects of our future, and we can create, through silence and thought, an inner and outer reality filled with harmony, optimism, peace, happiness, joy, and love.

It is a question of overturning the Latin phrase *mens sana in corpore sano* (healthy mind in healthy body) to experience the exact opposite: *corpus sanum in mente sana* (healthy body in healthy mind). If you continue to have an "emotional and mental diet" based on resentment, anger, guilt, helplessness, anxiety, and obsessive thoughts, your body will suffer, losing its harmony and health, even if you are physically active and eat healthily. We are just starting to understand how to balance this extremely complex and interdependent system. The body is not separated from the mind, the emotional and relational sphere, and spiritual life. In this context, spirituality doesn't mean a religious belief but rather awareness of the meaning, goal, and purpose of life. It is a secular and universal approach to the most intimate and authentic existential sense that lies in each of us. It's inextricably linked to the sacredness and beauty of the miracle of life.

Health, well-being, fulfillment, quality of life and relationships, understanding one's purpose in life, and expressing high values are all closely related to the proper functioning and achievement of a perfect mind. We need to understand how a perfect mind functions and what states it goes through. We need to know how to regenerate and train it to positively manage stress; how to activate the creative and intuitive processes that generate harmony, health, and success in life; and what its impact is on our health.

CORPUS SANUM IN MENTE SANA

According to the ancient wisdom of the Vedic *Upanishads*, the nature of the universe is mental. This means that everything we observe and define as real is nothing but the product of our mind. In his book *From Animals to Gods*, the historian Yuval Noah Harari writes that the human being is the only animal who can talk about things that exist merely in their imagination. This mental imagination is what differentiates us from other animals. Concepts and ideas such as nation, people, homeland, and religion are simply imaginary inventions that have, over the millennia, not only changed the trajectory of the social order and global geopolitics but also profoundly impacted the entire planet, nature, and all other forms of life. Our mind and imagination give rise to ideas that shape what we call reality (but that is still a mental illusion). When they are believed to be real by a critical mass of people, they are able to move power and wealth, determine the fate of entire populations, and influence the outcome of wars and revolutions. Even the abstract concept of money, which led to capitalism, the market, credit, and debt, is nothing but a figment of the imagination. Some of these abstractions have been used to create social and religious institutions, needs, and wills to push us to consume one product over another. Think of how borders, nations, and people were born. *Homo sapiens* believes that things that are only in their imagination are real; choices, decisions, and behaviors are the product of these beliefs turned into reality. How many people have died in a war for believing in or being forced to obey an abstract and imaginary idea? Imagination can bring prosperity or famine, life or death, sickness or healing. We know, thanks to modern science, that our mind is able to switch off or activate gene expression, slowing down aging and reducing inflammation. Our mind's activity affects not only the external environment and the future of the planet but also the internal environment and health of our biological machine. These two levels are therefore intimately interconnected and everything that happens to us is influenced by the processes that take place in our mind.

If it's true—as the millenary knowledge of the Vedic *Upanishads* proclaims—that the universe as we perceive it is nothing more than an illusion of our mind, then two considerations are equally true. The first, being aware of our mind's functioning, is that we can freely choose which reality to create, both in our internal environment and the external one. The second is that we can, if we really want to, observe the nature of life without illusions from our mind. How does the world, the universe, and ourselves appear without the filter of the mind?

According to modern science, the essential condition for studying the mind is to reduce it to measurable physical, neuronal, electrochemical, and biological processes. But are the mind and brain the same? Is the mind the product of the brain? Or is there an extracerebral mental dimension?

In his book *Homo cerebralis*, Michael Hagner writes that the world appears to our mind not in its reality but instead in what the brain "transmits" to our ego, processing the information it receives from the sense organs within the different cerebral areas. According to this point of view, consciousness, mind, memory, emotions, will, and language are all brain activities, and we would be nothing more than "a shell of neurons" without it, as neuroscientist Francis Crick writes. Nothing exists "without my brain," declares the manifesto of eleven German neuroscientists. Despite the enormous amount of data, however, neuroscience has not yet understood how the "intangible immateriality" of mind and consciousness manifests itself from matter. The main question that emerges from these lines of research is whether the mind that studies itself is capable of understanding how it is "produced" by the brain since it's the mind itself that conducts the investigation. Neuroscientists are aware of the current limitations in the study of mind and consciousness, and believe that the human brain "can never fully explain its own operations."

According to neuroscience, the way consciousness arises from neuronal activity still remains a mystery. Equally mysterious is the fact that there is an explicit correspondence between each thought and its neuronal correlates. Ideas, moods, values, planning, and sensations: Does reducing all of this to physical-chemical activities of the cerebral matter necessarily imply that the mind and brain are the same thing? We know that the activity of the mind modifies the brain, gene expression, biology, and even external experiences. Although we all have the same brain systems with more or less the same number of neurons, the way these are connected is different for each individual, and this diversity is what makes us unique.

If the brain creates the world we live in, is it possible to clarify in physical-chemical-neural terms what the mind's nature is and what it means to be "conscious"? What is self-awareness? Is the mind really just an electrochemical reaction? These questions should be addressed through a multidisciplinary approach.

The word *mind* commonly means a complex of cognitive activities and emotional states that do not always and necessarily translate into behavior and action. In Western culture, the prevailing view is that humans are able to distinguish themselves from the rest of the animal kingdom because they have the exclusive ability to think. Human beings have felt superior for "only they" having a mind.

In this regard, scientist and philosopher Blaise Pascal writes, "Man is but a reed, the most feeble thing in nature; but he is a thinking reed. The entire universe need not arm itself to crush him. A vapor, a drop of water suffices to kill him. But, if the universe were to crush him, man would still be more noble than that which killed him, because he knows that he dies and the advantage which the universe has over him; the universe knows nothing of this. All our dignity consists, then, in thought."

On the contrary, Indo-Vedic culture has always viewed the mind as a potential obstacle to the realization of the intimate, authentic, and innate nature of the human soul. This culture considers the mind a limited and apparent tool, and its "extinction or overcoming" is seen as a fundamental step in canceling human ego. The person's growth, in this perspective, passes precisely through understanding the real nature of the mind and overcoming it.

Human beings greatly resist the idea of being part of a unitary reality and being interconnected on infinite levels with everything surrounding them. This resistance, this sense of separation and superiority, originates from the idea of an "individual self," a "personal mind," a "feeling, thinking, and working ego" separated from anything else. The maximum expression of this self is personal free will. This "egocentric" perceptive and cognitive distortion produced the anthropocentric evolutionary model and generated the geocentric vision of the universe.

"A human being is a part of the whole, called by us 'Universe,'" said Albert Einstein, "a part limited in time and space. He experiences himself, his thoughts and feeling as something separated from the rest—a kind of optical delusion of his consciousness. This delusion is a kind of prison for us, restricting us to our personal desires and to affection for a few persons nearest to us. Our task must be to free ourselves from this prison by widening our circle of compassion to embrace all living creatures and the whole of nature in its beauty."

THE PERCEPTION OF REALITY

On the evening of October 30, 1938, the CBS Radio building in New York City was under investigation. An army of people in uniform searched every corner to seize copies of recordings of a show that aired right before. A few minutes later, the press made an abrupt entrance, attacking actors, supervisors, and producers in search of firsthand news. What did they know about the panic, suicides, and fatal accidents and stampedes on the street in every part of the country? The show's lead actor and creator, a twenty-three-year-old man, sat to the side looking dejected and resigned to the imminent failure of his career. He was unaware that things would turn out quite differently for him, resulting in what would be one of his greatest masterpieces, *Citizen Kane*. That sad, unconsolable boy was Orson Welles. But what had happened a few hours earlier?

Like every Sunday, the *Mercury Theater on Air* aired, a niche program without sponsors or advertising. It was Halloween, and Welles literally adopted the classic children's phrase "Trick or treat? but choose the first option: trick. Thus by readapting the novel *The War of the Worlds* by his namesake, H. G. Wells, he moved the setting from the Victorian suburbs of London to the New York area. He made the story sound like a live chronicle of a Martian landing and attack, complete with interviews by experts, official bulletins, statements from the authorities, eyewitnesses, and sound effects of explosions, sirens, and screams of the victims. It didn't take long to realize that the prank had turned out all too well. A third of the six million listeners who gradually joined the program through word of mouth believed what they were hearing was real. The result was shocking: traffic jams caused by citizens pouring into the streets en masse trying to flee, police switchboards overloaded with hundreds of telephone calls, and a rampant wave of panic. The joke of the Martian invasion had become a reality, with all of its side effects. The phantom aliens had truly conquered—if not the earth—the minds, thoughts, and reactions of millions of people.

We often underestimate the power of imagination, thought, and words, but they can move the masses, regardless of the historical moment, geography, and social and educational status. History teaches us that what really matters are not the things themselves but instead how we perceive them. Let's think of how Adolf Hitler managed to shape the minds of millions of people under fascism, or the power of Gandhi's thought. Or more simply, how advertising influences the entire economy every day, generating desires based on emotional associations with products that most of the time, we don't even need. In an increasingly interconnected world, we can no longer avoid becoming aware of the mechanisms that guide and bind thoughts, words, and actions. The human brain is inundated with an enormous amount of information that would cause even a powerful computer to crash.

A study conducted in 2008 at the University of California at San Diego found that on average, people are inundated every day with information equivalent to 34 gigabytes, enough information to overload a laptop within a week. This number includes receiving about 100,500 words daily (23 words per second factoring in sleep time). Are we biologically and mentally ready to support this type of stimulation? What changes and adaptation will it require? In this time of information overload, fake news is on the rise. Ad hoc modified algorithms bias news reporting, and the media compete for sensational headlines. Can these dynamics further increase the disconnection between "objective" data and the reality we perceive? To answer this question, Owen Shen and his team, in a recent analysis of Our World in Data, compared four key sources of data related to the causes of deaths in the United States, highlighting some staggering results.

About a third of all deaths are caused by heart disease, yet this receives only 2–3 percent of Google searches and media coverage. A little less than a third of all deaths result from cancer, a figure similar to Google search statistics (37 percent of searches), but this only receives 13–14 percent of the media coverage. Suicide, homicide, and terrorism get far more attention in Google searches and the press than their actual percentages. In terms of media coverage of causes of death, violent deaths account for more than two-thirds of the coverage in the *New York Times* and the *Guardian*, but less than 3 percent of the total deaths in the United States.

An interesting result is that the percentages of searches on Google are much closer to real data than the attention given by the press. This reflects the discrepancy among three elements: the perception created by the press, real facts, and information when it's searched independently.

Just as when working on perception and mind, it is possible to create phobias, panic, and ideologies directing crowds toward a particular goal, it is possible to rebalance the entire system as well and make conscious choices. Our responsibility lies in the awareness of how the information we receive can transform our destiny. We need to unravel the instinctive and unconscious reaction triggered by leveraging our primal survival instincts in order to defuse these hardwired responses. Educating our mind to kindness, harmony, and beauty is possible, if we know the nature of the mind and how it functions.

EDUCATE THE MIND

PURITY, PRESENCE, AND INTEGRITY

Usually the mind is engaged in a vortex of endless analysis, processing stimuli coming from external and internal environments. It's a constant inner dialogue based on thoughts, reflections, judgments, interpretations, and impressions that exhausts us and makes us lose vital energy. So the first step toward a perfect mind is understanding the states it goes through, becoming aware of its characteristics, and finally regulating its functioning.

If not correctly trained, the mind of a person tends to be extroverted and scattered; it directs its attention to external objects and situations, and unconsciously reacts to emotional wounds (betrayal, injustice, or abandonment), fears, and anxieties for the future or traumas of the past in an endless rumination.

A perfect mind is purified of these tendencies and is whole. It's a meditative mind trained through silence, harmonic creativity, and intuition, and it's intensely immersed in the present.

If you happen to have a dog, you can understand how important it is to educate them correctly. An untrained dog can complicate our lives as well as destroy our house. An untrained mind generates suffering and conflict, is distracted by memories of the past or creative projections of the future, and influenced by judgments, fears, and anxieties. The mind repeats cyclic patterns that make the same mistakes, choices, and situations over and over again. It has obsessive and recurring thoughts, and is hypercritical, intolerant, and full of prejudices. It expresses a sense of superiority or inferiority.

We can define it as the "unaware reactive mind" because it is unknowingly at the mercy of endless stimuli and becomes the product of them.

When the mind is trained through a healthy lifestyle, discipline, and meditation, it becomes a wholesome tool; it can conceive, create, and see beauty and harmony everywhere, even when in the deepest pain and discomfort. The mind is no longer the product of past ghosts and fears for the future, but captures and interacts with life in the present, just like children do. It finds inspiration through the contemplation of silence, emptiness, and infinity. It is able to reason unconditionally, without being influenced by the logic of convenience or expectations; it doesn't react impulsively to stimuli. It replaces the sense of superiority or inferiority with an awareness of uniqueness and complementarity.

The perfect mind is introverted: it sees the world and interacts with it, but it does so from within and doesn't get lost in it. It is able to clearly analyze its inner states and recognize them, and has the ability to control, transform, and create its own internal environment independently of the external environment. This means that it's no longer what happens outside that influences our inner state, but on the contrary, it's our inner state of clarity, inspiration, and kindness that deeply affects and influences the external environment. The external reality completely depends on the internal state. The more the mind internalizes and seeks the origin of the world within itself, the more we are aware that each of us is the author of our own well-being and quality of life. In this kind of mind, a new and deeper sense of responsibility arises.

US psychologist and philosopher William James wrote that many people think they think, but in reality they are just rearranging their prejudices. The process of harmonizing the mind doesn't consist in controlling its activity but rather in completely resetting it. One can block its dispersive tendencies and bring it back to an original condition of purity, clarity, and full awareness. In this sense, practicing meditation plays a fundamental role. The perfect mind is as pure as that of a newborn child. It is crystalline and not crowded with thoughts, judgments, criticisms, disappointments. It shines in the present and is not projected onto a past that no longer exists nor into a future that doesn't exist yet.

Adults have lost their ability to feel the harmony of the present moment—the here and now. In the present moment, if it's lived with awareness, we can experience happiness and perfection. Our mind replaces the wonder and beauty of life's miracle, perfect in all of its manifestations, with its own artificial projections. For children, everything is different. They have absolute natural access to the perfect moment—this instant. The younger the child, the closer to a state of purity and integrity, such that they don't understand what "the day after tomorrow," "tonight," or "one hour" means.

THE SEVEN ALARM CLOCKS EXERCISE

A good exercise to test and improve your state of mindfulness is to set seven alarm clocks throughout the day. Each time the timer chimes, the challenge consists in stopping it, taking three deep breaths, making contact with the present moment, and mentally repeating this phrase seven times: "I am totally here. Now." The seven alarm clocks exercise has several functions, the first of which is to train the mind to bring itself back to the present time. But it also improves clarity and mental focus, promotes relaxation, and allows you to better manage stress.

LIVING IN THE FLOW

The "state of flow" is a feeling of freedom derived from the absence of negative thoughts coupled with a particular condition of self-awareness. The state of flow was theorized in 1975 by US psychologist Mihály Csíkszentmihályi, who observed that under certain conditions, people persist in what they do, ignoring any discomfort or tiredness, if the act produces in them an inherent satisfaction. The state of flow produces a strong intrinsic self-motivation (autotelic)—that is, an intense satisfaction and enjoyment in what one does, regardless of the final result one obtains. It's an experience of complete and total immersion in the task being performed. Being able to enter the state of flow depends on how the person perceives the challenges that the external environment offers them and what skills they can count on to deal with them. The balance between skill and perception of the challenges will lead to the experience of flow: a total involvement (behavioral, cognitive, and emotional) of the person, for a pure and complete enjoyment in a state of absolute presence.

In sporting circles, there is increasing awareness of the importance of the mind in achieving results. There are even those who are convinced that the success in many sport competitions is mostly determined by how correctly one uses the mind. A lack of focused attention and moment of distrust can be fatal for the race to the podium. A saying attributed to boxer and activist Muhammad Ali is that "If my mind can conceive it, and my heart can believe it—then I can achieve it." Roger Federer lost the tennis semifinal against Novak Djokovic during the 2011 US Open by missing two match points on a serve in the last set, when everything was going well just before. Sometimes even a single thought or a dysfunctional mental state, not only in sports, but in dangerous situations involving the police, a firefighter, doctor, or nurse, can be the difference between success and failure. Being aware, fully present, and in a state of flow during these moments is the key to success. For this reason, more and more professionals are turning to meditation to prepare their mind and improve focus and attention as well as reduce stress and anxiety.

AN INSPIRED MIND

Johann Wolfgang von Goethe said, "One ought, every day at least, to hear a little song, read a good poem, see a fine picture, and, if it were possible, to speak a few reasonable words."

According to the ancient Greeks, an inspired poet would get in touch with the thoughts of the gods and fall into ecstasy, which would have transported them out of their mind. The word *inspiration* comes from the Latin *inspirare—that is,* "to blow in," to ignite a deep feeling or elevated idea in someone's soul. An inspired mind is available and open to infinity. This allows it to be fertilized by feelings of a higher order and sublime ideas; it's the flame of something infinitely great that is blown into us.

What are the contexts, people, and situations that inspire us?

Through which sensations can we define and recognize that we feel inspired?

What does being inspired mean to you?

CULTIVATE INSPIRATIONAL HABITS

Make a list of seven things (experiences, actions, people, and situations) that truly inspire you and turn them into daily rituals.

INNATE TALENTS

Have you ever heard Lang Lang, one of the world's most famous and virtuosic pianists, play Franz Schubert's *Military March* with five-year-old prodigy

Ricky Kam? Other prodigy pianists such as Umi Garrett, Ethan Bortnick, and Alexandra Dovgan are also incredible examples of how talent manifests itself through innate tendencies from an early age.

It is fascinating to see how this phenomenon is explained by the millenary Indo-Vedic culture through the nature of mind. Such latent innate tendencies, both positive and negative, present in each of us from birth, can also manifest and flourish suddenly and unexpectedly in certain conditions and moments of life. They can be destructive talents and impulses, unconscious traits of the character, or virtues.

Perhaps the old adage "If you sleep with dogs, you wake up with fleas" is true? It depends. It depends on the innate tendencies of each individual. On the other hand, lotus flowers grow right in the mud, and there are thousands of stories of people who grew up in violent and challenging environments, yet were not affected by them. They followed innate predispositions and instincts that allowed them to grow anyway in virtue and integrity.

Places, people, or situations are not ultimately responsible for what we think, feel, or do, but we are. If there's only peace in us, only peace will be expressed, regardless of the external circumstances. Our intimate feelings constantly draw the line of our destiny. It all depends on what we feel we are; it depends on the innermost thoughts we feed and deep latent tendencies that unknowingly exist within us. What do these tendencies depend on? How were they formed and where did they come from?

The mind, in Patañjali's yoga sutra, is indicated by the Sanskrit term *manas* and considered a "subtle organ endowed with awareness." Unlike Western culture, which perceives the mind as something abstract, a nondefinable place where our thoughts come to life, for Indo-Vedic culture it is a real organ, just like the heart, lungs, liver, and kidneys. Unlike these, however, the mind is considered an organ composed of matter defined as "subtle," with the function of processing and storing thoughts, ideas, and impressions. It is a database in which an extraordinary wealth of information is stored in a latent state. Again, according to this fascinating tradition, the mind survives the physical death of the individual, carrying with it—in the course of successive existences—all the latent memories (defined by the term *samskara*) present in a potential state, but ready to express themselves under certain conditions. They influence innate talents, tendencies, character, and behavior. Regardless of the social context in which we were born—the influences of parents, friends, and educators—there is therefore a potential latent tendency in each of us that depends precisely on the nature of our mind. This is why the ancient Eastern traditions have paid great attention to the health and purification of the mind, resorting to tools such as contemplation, inner silence, meditation, and breathing. This is intended to undo the unconscious destructive tendencies, considered the cause of our suffering. The key question here is, If modern science has shown that the mind can influence DNA, modifying aging and inflammatory processes, cognitive abilities, memory, mood, anxiety, and depression, what influences the mind? What tools can we use to manage it, purify it, and make it a perfectly functional organ?

THE THREE QUALITIES OF THE MIND

How can one have an enlightened, inspired, forever young mind?

Nature, according to Indo-Vedic culture (*prakti*, meaning "immanent reality" or "manifest nature"), is characterized and influenced by three qualities called *guna* (a Sanskrit word that indicates "virtue" or "attribute" too).

The three gunas, considered the ultimate components of matter, are:

- *Tamas*, "darkness," the heavy forces of inertia that hinder dynamism
- *Rajas*, "dynamic," indicating the activating component of life
- *Sattva*, "truth" or "that which is," indicating the forces that uplift and illuminate

Each of these three qualities is neutral, neither positive nor negative. The quantity, in excess or deficiency, leads to balance or disharmony. An excess of tamas generates heaviness, inertia, numbness, and drowsiness, while its deficiency generates insomnia and difficulty getting full rest and regeneration. Tamas, when in perfect balance, guarantees a healthy rest. An excess of rajas causes hyperactivity and dependence on doing, while a deficiency causes a lack of concrete action and dynamism along with excessive static in every area of existence. Rajas, when in balance, guarantee the right dynamism.

An excess of sattva can cause one to be scattered and lack concreteness as well as pragmatism—not having your feet on the ground and instead having your head in the clouds. A deficiency of sattva can cause a lack of inspiration, excessive earthliness, and venality. Sattva, when in balance, guarantees the right carefreeness and inspiration.

The three gunas are found in every aspect of existence and hence influence the activity of the mind (manas).

The tamasic mind is when tamas prevails. The mind is immersed in numbness, sloth, apathy, heaviness, and ignorance. Tamasic individuals tend toward laziness and inertia.

The rajasic mind is when rajas prevails. The mind is in a state of instability, hyperactivity, constant desire, and restlessness. Rajasic individuals tend to be emotional and passionate, and immerse themselves in doing and seeking immediate gratification.

The sattvic mind is when sattva prevails. The mind is in a state of clarity, wisdom, purity, virtuosity, and serenity. The presence of sattvic individuals elevates, bringing lightness, peace, balance, and inspiration.

WHAT KIND OF MIND DO YOU HAVE AND UNDER WHICH CIRCUMSTANCES: TAMASIC, RAJASIC, OR SATTVIC?

Which sensations, among those listed in the table, do you feel more often during the day?

What contexts, experiences, and people are they associated with?

Think about a person, situation, or event in your life and how it makes you feel. Does heaviness, dynamism, or lightness prevail?

Tamas	Rajas	Sattva
Relaxation	Dynamism	Equilibrium
Inertia	Frenzy	Harmony
Inactivity	Energy	Positivity
Negativity	Excitement	Peace
Apathy	Focus on doing	Clarity
Boredom	and the result	Lightness
Heaviness	Desire	Creativity
Laziness	Worry	Openness
Disinterest	Restlessness	Serenity
Lethargy	Anxiety	

A QUESTION OF BALANCE

Imagine the three gunas as three lakes:

- The first, tamas, is full of mud so the water is murky. You can't see anything in the muddy water. Even if it's a bright and sunny day, you won't see any reflections. This is why tamas is also known as the dark guna.

- The second, rajas, is agitated so the water is always in motion. There may be some light reflected off the surface of the water, but it appears only in brief, fragmented glimmers. We can see that there is light, but because of the agitated water we believe that the light is also moving.

- The third, sattva, is completely still, limpid, and clear. There's no murkiness or agitation; light shines through the water. That's why sattva is considered "the revealing truth." When the mind is sattvic, we see things as they really are.

REBALANCE THE TAMASIC TENDENCY

If you find yourself overwhelmed by tamas, you will likely feel lazy, confused, and physically, mentally, and emotionally heavy. Doing anything, even just thinking, can be exhausting. Tamasic states of mind include fear, depression, nihilism, discouragement, and a lack of motivation. Here are some ways to rebalance the tamasic state:

- Movement: reactivate yourself through exercise, training, walking, or running. Moving the body moves all of your energy, which in turn stimulates a change in your thinking and emotional patterns. Exercise is probably the fastest and most efficient way to manage tamasic states.

- Micro objectives: the tamasic tendency generates powerlessness and discouragement. We can reverse this mindset by adopting a more proactive approach to what makes us worry. If something doesn't work in our life, we pragmatically decide how to improve it, and start setting ourselves some short- and medium-term goals. Even small steps forward can make us feel so much better.

- Avoid junk food, alcohol, and psychoactive substances: the devastating effect these have has on the mind and body is not worth the momentary pleasure.

- Select stimuli: pay attention to how you use media. What do you watch on TV every day? What appears prominently on your social media feed?

What kinds of movies do you watch? Reduce or eliminate anything that contributes to a tamasic mindset.

- Leisure: Listen to vibrant, light music; choose readings that inspire and motivate you.
- Relationships: Pay attention to who you spend time with, reflect on the effect that certain people and relationships have on your mind and emotions, and spend more time with those who can inspire and energize you.

REBALANCE THE RAJASIC TENDENCY

If you find yourself experiencing a hyperactive, restless, anxious state, and your mind is constantly ruminating, full of judgments and thoughts; if you can't stop and always need to do something; if you struggle to fall asleep or can't sleep well—these are all indications that you must rebalance your rajasic state. Here are some ideas:

- Meditate: meditation is perhaps the greatest balancing activity. It is effective in calming the mind and body. A large amount of studies have clearly shown that meditation restructures the brain, with many long-lasting benefits (see chapter 13, "Meditate, People, Meditate," p. 155).
- Spend time in nature: even a short walk outdoors in the woods or on the beach can harmonize the mind and body. Nature is extraordinarily balancing and healing (see chapter 17, "Nature Heals," p. 197).
- Nutrition: reduce or eliminate as many stimulants as possible such as caffeine. Try replacing your coffee with a glass of water with lemon in it or herbal tea.
- Tech detox: take a break from media (your TV, computer, social media, or smartphone). Go offline for a day or two and notice how this improves your mood. Alternatively, limit the amount of time you spend using your phone, social media, or email.
- Select stimuli: pay attention to the news you listen to, watch, and read. The media often captivates the audience thanks to catastrophic, sensational news with great anxiety-producing potential. Consciously select the incoming information flow.
- Create a new playlist: reduce the music that increases your excitement and frenzy. Prefer music that calms and rebalances you (see the recommended songs listed at the end of chapter 15, "Music and Sound: Health, Well-Being, and Longevity," p. 188).

- Rest and sleep: learn how to consciously relax. You can do this through the simplest tool you have in every moment: conscious breathing. The physical and psychological benefits are immediate and profound throughout the psychophysical system (see the "A Strategy to Calm the Mind" section of chapter 12, "The Power of the Mind," p. 152).

- Relationships: pay attention to the people you spend your time with, reflect on the effect certain people and relationships have on your mind and emotions, and prefer relationships that make you feel calm and happy.

- Enjoy the ride: instead of seeing everything as a desperate race, learn to enjoy the ride. Gratitude is one of the main tools for savoring every moment (see chapter 4, "Forgiveness and Gratitude," p. 43).

THE FIVE STATES OF THE MIND

Can you recognize your most frequent mental state?

To develop an enlightened mind, you need to be aware of the five states it goes through.

The first is called *mudha*. This state is dominated by the tamas guna, and produces a mental state of numbness, confusion, heaviness, tiredness, apathy, and doubt. You feel clouded, with a tendency to apathy and inertia. The mind is attracted to anything that creates heaviness and numbness, such as hours spent in front of a TV or computer, fatty and ultraprocessed foods, demanding relationships, and drug or alcohol abuse.

The second is called *ksipta*. It's produced by the rajas guna, which drives the mind to be restless, hyperactive, totally extroverted, and attracted by sensory pleasure. The mind finds pleasure in being hyperstimulated, experiencing states of excitement and euphoria in which one has the sensation of "going to the maximum," pressing the accelerator, and feeling "pumped up." It is drawn to shapes, colors, relationships, and emotions that create intense sensory pleasure.

The third is *viksipta*. Mental activity begins to be influenced by the sattva guna, but in an unstable way. This instability tends to create a distracted, superficial mind; one has the sensation of being "up in the air," not concrete, or with one's head in the clouds. The attention fails to focus, except on pleasures, carefully avoiding problems, responsibilities, and anything that could create worry, anxiety, apprehension, and suffering.

The fourth is *ekagra*. This state comes from the sattva guna too, but unlike the previous one it is balanced. The mind is stable, steadfast, and has clarity and presence; it manages to stay focused. It's a mind willing to be inspired as well as to create, nurture, and cultivate high thoughts. It's a good condition to meditate.

The fifth and final state is called *niruddha*. The mind is devoid of thoughts; the constant internal dialogue made up of images, ideas, and impressions is suspended and replaced by mental silence. Attention no longer contemplates any object. All mental activity is inhibited. It is a condition of deep regeneration. This state is the result of a particular meditative practice.

Of these five states or modalities of the mind, the first three are not suitable for the practice of meditation. If you try to meditate without first having regulated and "tamed" the mind, it will be impossible, inaccessible, and incomprehensible as well as frustrating. The mind will resemble an uneducated dog, unable to control themselves, and prey to instincts and impulses. It will be distracted by the solicitations of the surrounding environment and thus unable to respond in a balanced way. The last two states (ekagra and niruddha) are both suitable for meditative practices.

The three characteristics of the mind (tamas, rajas, and sattva) and the five mental states that derive from them (mudha, ksipta, viksipta, ekagra, and niruddha) are influenced and regulated by eight pillars of our life:

1. Healthy eating
2. Regular physical activity
3. Healthy and happy relationships
4. The regular practice of meditation
5. A harmonious relationship with nature
6. A healthy and balanced sexuality
7. Listening to suitable music
8. Reading inspirational texts

At the end of this book, you will find a chapter entirely dedicated to the guidelines for healthy eating, physical activity, and meditation for different age groups. You will also find reflections, advice, and guidance in the chapters dedicated to healthy, happy relationships and music as well as inspired reading tips to feed the mind in the right way.

WHAT ARE YOU FEEDING YOURSELF?

Tamasic Food	Rajasic Food	Sattvic Food
Garlic	Mustard	Fresh fruits
Margarine	Black pepper	Fresh vegetables
Hydrogenated Fats	Asafetida	Cereals
Milk chocolate	Horseradish	Legumes
Seitan	Seaweed	Dairy from nonintensive farming
Tofu	Umeboshi	
Glutamate	Radishes	Dried fruits
Mushrooms	Carrots	Oilseeds
Potatoes	Eggplants	Pumpkin seeds
Alcohol	Pineapple	Oven, crock, pan, and spit cooking
Vinegar	Fructose	
Refined foods	Chestnuts	
Processed foods	Myrtle	
	Tomato sauce	
	Spicy sauces	
	Lactobacillus	
	Probiotics	
	Caffeine	

A STRATEGY TO CALM THE MIND

The close relationship between mental activity and breathing has been highlighted by hundreds of scientific articles. Just go to PubMed, the most important database of scientific articles in the world, and type "breath + respiration + brain + meditation" to get confirmation. What has been validated by science in recent times was already affirmed and explained in thousand-year-old texts such as the Vedas that consider breath "the densest part of the mind," affecting its activity and states. Relaxation and mental calm can therefore be achieved through breath control and conscious breathing. In particular, these ancient traditions emphasize the regulation of expiration and retention of breath with empty lungs. Breath suspension or retention performed with empty lungs (at the end of the exhalation) is called *rechaka*, while breath retention performed with full lungs (at the end of the inhalation) is called *puraka*. The goal of this type of breathing is to calm the mind, which is no longer at the mercy of a flurry of thoughts, but enters a state of clarity, concentration, and focus, ideal for meditation.

In Patañjali's yoga sutra, we speak exclusively of breath retention with empty lungs (rechaka) because we want to create a sort of mental vacuum, favored by the physiological effect of airless lungs. During full lung air retention (puraka), the body is more energized, so the mind can easily focus on a certain object of attention.

BREATH OVERCOMING THE MIND

Sit in a comfortable and relaxed position.

Pay attention to your breath going in and out of your nostrils, and become fully aware of it.

Take twenty-one breaths, progressively deeper, natural, relaxed, and slow. Intentionally perform each exhale to completely empty your mind.

Every three breaths, empty your lungs completely and hold your breath, remaining in suspension for at least seven seconds. The retention time must be established on the basis of listening and self-awareness, without forcing it. The goal is kindness and mental calmness.

While holding your breath with empty lungs, explore and listen to the void that is created.

The first breath after each retention will be deeper and wider.

At the end of the third and final retention, remain silent for a minute, with a calm mind, open to listening, being present, and remaining aware.

13 MEDITATE, PEOPLE, MEDITATE

DANIEL LUMERA

> Ours is a fortunate condition. We are animals with one foot in infinity.
> —Daniel Lumera

Today meditation is one of the strongest trends. Along with healthy nutrition and physical activity, it has become one of the three pillars of well-being and quality of life. It's the fastest-growing health care practice in the United States, according to the US Public Health Centers for Disease Control and Prevention, with the number of meditators tripling between 2012 and 2017. More than eight thousand scientific studies have been conducted on its effects, with extraordinary results in the fields of medicine, psychology, and neuroscience. Meditation, however, is being commercialized in the Western world, where its value is considered transactional. The focus is now on the extraordinary benefits it produces: offering relaxation, fighting inflammation, slowing aging, regulating mood, reducing depression, and improving cognitive abilities. Although belonging to the sphere of "knowing how to be," meditation is appreciated especially for its effects. It has therefore been moved to the area of "knowing how to do" as a skill to be acquired in order to get lasting well-being and health. But meditation is much, much more than that.

Its success in recent years is basically due to four factors. The first is its positive impact on health and well-being. The second is the need for balance given fast-paced, noisy, stressful, and competitive lifestyles. Haste, hyperactivity, and emotional roller coasters require experiences that restore calm, listening, internalization, peace, and silence. The third factor is evolutionary, related to an awareness of the meaning, role, and purpose of life.

Human beings must reconnect harmoniously with the environment, all other forms of life, and the universe, transforming themselves from exploiters of resources and other beings into part and parcel of a delicate balance. Meditation in this sense allows one to increase awareness and expand one's consciousness, guaranteeing a much broader vision of life and oneself. The fourth factor is existential. It regards realizing the most intimate, elevated, and essential part of oneself. It is call to reach higher states of consciousness where one experiences the nature of happiness, and returns to an original purity and integrity.

Meditation is not just a fad; its true essence has ancient roots and deep meaning. Today we meditate in companies, squares, public gardens, offices, and churches. In a time where there are more holistic centers than bank branches, it is good to ask yourself what meditation is really about.

THE REVOLUTION OF BEING

My parents used to see me sitting in meditation for hours and didn't quite understand what was going on. In a culture of doing and thinking, perhaps remaining motionless in silence and awareness of being can seem like a waste of useful as well as precious time. But they didn't know that in my brain, millions of structural transformations were taking place—something that today's science is starting to explain. From the ability to slow down aging, develop memory, concentration, and attention, reduce inflammation, activate and deactivate gene expression, and regulate mood, including the possibility of modifying the DNA, meditation has an enormous impact on the health of body, mind, and spirit, and on the quality of individual, relational, and social life.

The evolutionary reasons in that call were huge, not only related to well-being, but to a new way of being human, of feeling oneself, life, and nature. The same happened when the generation of peasants' children asked to study. Many were denied that possibility and were sent to the fields to cultivate because "knowing how to think" would not have brought bread to their table, and the best investment was "knowing how to do" in the primary good of the earth. Yet the thought would turn out to be far less abstract than what they believed in creating and shaping reality. Today the same thing is happening. There are more and more people who have understood the importance of knowing how to be. There are far more people who recognize the significance of knowing how to do, think, and be. The value of awareness will soon pay off in the lives of those who welcome this inner revolution in terms of health, well-being, prosperity, and harmony with life. These are all evolutionary necessities of fundamental importance not only to live but above all to survive on this planet. Empathy, kindness, forgiveness, and awareness—this is the food of the near future.

They used to watch me sitting in silence, motionless, even for hours, and perhaps they didn't understand that I was passing from "knowing how to do" and "knowing how to think" to "knowing how to be."

ONE WORD, ONE WORLD

Meditate, medicate, and *medicine* share the etymological root *med-,* which derives from the Latin *mederi* and means "to cure." Meditate thus derives from the Latin *meditari* (iterative of *memeri,* "cure"), and in its original meaning, it meant "practicing" and then, subsequently, "reflecting," elaborating and preparing in the mind something we intend to achieve. Meditating has always indicated the mind's ability to ponder and dwell long and carefully on an object of attention (an idea, topic, text, or problem) to investigate and understand it. For example, meditating on a passage from the Bible is not simply a logical or superficial reflection but instead a deep and conscious listening. In Eastern philosophy it indicates a process of realization aimed at self-awareness, which is obtained when one greatly masters the activities of the mind in order to refine its ability to fix attention and concentration on a single thought, excluding all others. This focus on a precise element of reality (or an elevated thought) has the initial objective of stopping the usual flow of thoughts (dialogue and background chatter). The mind becomes absolutely silent, still, and peaceful, resting in a natural state without activity or disturbance.

Meditation is recognized and practiced in different forms by all major religious traditions. In the *Upanishads,* there is the first explicit reference to meditation that has come down to us: the Sanskrit term *dhyana* (literally "vision"). There is no word in any other language that has the same specific meaning for the simple fact that it indicates a state of being that was not clearly identified. For millennia it remained without translation because the state of consciousness that it denotes was not recognized with certainty.

The meaning that is commonly given to the term *meditation* is "concentration" or "attention," two elements that are part of the process through which one reaches the "meditative state." When we confuse the meaning, the word meditation implies an object of attention: "What are you meditating on?" "I meditate on truth, I meditate on beauty, I meditate on silence, I meditate on the present, I meditate on God." According to this perspective, it is paradoxical to simply say, "I am meditating," because the statement would be incomplete. But dhyana is not simple attention, concentration, or contemplation; "being in meditation" contains its whole meaning. Forgive the apparent paradox, but by the word meditation in this book, we don't mean it as it is commonly understood but rather as the state of mind defined as dhyana in the *Upanishads.*

Two thousand years ago, the same problem was faced in China and Japan when the Buddha used the word *jhana* to refer to specific states of mind. That word was the transliteration in Pali (the Indian language of the Indo-European family; it is used still today as a liturgical language in Theravada Buddhism) of the Sanskrit term dhyana. Buddhist monks could not find a single word to translate this term: jhana therefore arrived in China, becoming *chán*, and then in Japan, *zen.*

The Chinese chán is nothing more than an attempt to imitate the sound of the Sanskrit dhyana and represent it with an ideogram indicating the sound and not the meaning. Literally chán means "flattened altar" or also "to abdicate." But although this ideogram was chosen to imitate the sound of dhyana, it is made up of two ideograms: the first means "to indicate, to point to, to show," and the second means "only, unique, simple, single." Therefore, separating the meaning of the two parts, it turns out that the ideogram contains the meaning of "aiming for simplicity" or "aiming for the one."

WHAT IT IS AND WHAT IT'S NOT

Meditating is not visualizing. "Close your eyes and imagine walking on a wonderful beach." These mental images and the whole visualization process have nothing to do with meditation.

Meditation is not conscious breathing. The breathing practiced to achieve a state of conscious presence can in no way be defined as meditation. Conscious breathing was considered by ancient wisdom philosophies as one of the preparatory elements. So the practice of conscious presence doesn't consist, as it's often and erroneously believed, in the deep meditative state of consciousness indicated in the Indo-Vedic tradition. The processes of attention, concentration, and full awareness of a state of presence are related to the first steps to be taken in order to access the meditative state.

Meditating is not a relaxation process. The practices of relaxation and release of tension and stress are not considered meditation.

In the West, "I meditate on it" means starting a reflective analytic process to metabolize and process information. Meditating is thus equivalent to thinking. Nothing is further from the true meditative experience, which functions to precisely cancel any mental activity. Meditating, then, is not thinking, reflecting, or elaborating.

Meditating is not praying. Many Eastern teachers have emphasized the curious statement, "Praying is when you speak and God listens. Meditating is when God speaks and you listen."

Meditation shouldn't even be confused with a new age or esoteric practice because it belongs to various traditional wise philosophies, including the Vedantic, Taoist, and Buddhist trends, among many others. Associating or attributing it to recent movements is a common mistake.

You can also hear people say, "He went into meditation and started channeling." It's good to definitively debunk certain myths. Meditating is not about channeling anything. You don't receive any type of information, message, or image. In the meditative state of consciousness, all visual and perceptive phenomena are exhausted. It doesn't consist at all in seeing something particular, or feeling particular emotions or sensations (such as joy, a sense of union, or expansion), and you don't experience visualizing lights, colors, or images. None of this is meditating.

Probably at this point many of you are wondering, "So what is meditation?"

Meditation is a state of consciousness that occurs when only pure awareness of being remains in the mind, without forms, definitions, names, judgments, or levels of identification. It's just a pure, infinite ocean of awareness. This state is reached through four distinct and simple phases with the guidance of an expert, discipline, and a correct lifestyle. It is within everyone's reach.

FOUR SIMPLE STEPS

How do you get to the state of meditation?

To experience the meditative state, the mind, as we have already said, must pass through four distinct and specific phases: focused attention, sustained concentration, deep contemplation, and finally, the state of meditation.

FOCUSED ATTENTION

Usually our attention is scattered or otherwise not used to focus on a single object without getting distracted. There are infinite attractions and stimuli (both internal and external) that keep the mind engaged in a constant internal dialogue. Let's imagine a viewfinder used to focus on a single object of attention; for example, it could be the image of the sun, the point between the eyebrows at the root of the nose, or even an idea, image, emptiness, or silence. The object of attention we choose should be something that inspires us, that brings to mind impressions of peace, serenity or strength, well-being, and clarity. When attention is totally focused on it for a long time, then we enter the stage of sustained concentration.

SUSTAINED CONCENTRATION

Sustained concentration has three characteristics: it is stable, intense, and absolute. It doesn't involve effort but rather presence; otherwise, it would generate tension and fatigue. It's a state of intentionality and will: we are active in supporting it, but without any kind of forcing. On the contrary, we are completely relaxed, present, and comfortable in this dimension of centering. In the state of sustained concentration, there is a perfect balance between the electric and magnetic components.

The electric characteristics are determination, will, focus, and proactivity. The magnetic characteristics are availability, trust, and knowing how to abandon and let go. When this balance reaches maturity, and is effortlessly prolonged and sustained over time, we naturally enter the stage of deep contemplation.

DEEP CONTEMPLATION

This stage is characterized by three elements: mental silence, no definition, and no judgment. Watching a sunset or contemplating it are two completely different things. In observation there is interpretation and definition, and there can be judgment: "Look at the beautiful sunset, look at the beautiful colors."

In contemplation, on the other hand, one observes through a state of mental silence—without thoughts, considerations, prejudices, and judgments. The mind becomes like that of a child's, observing things for the first time without tracing them back to something known—a state of wonder and nondefinition. No names or labels are attached. The mind returns to an original state of purity. If we contemplated the world all the time, our mind would live in a state of perpetual wonder and excitement. When deep contemplation is sustained and prolonged, we naturally enter the meditative phase.

STATE OF MEDITATION

In this phase, a condition of calm and lightness is reached, in which the breath is deep. There is no longer the feeling of "performing a practice" and something is highlighted that previously slipped away but has always been there: a sense of integrity and shining purity, a prelude to a unique sensation, as if the soul was smiling. One must learn to let that state take over and surrender to it. The meditative state contains all the ingredients of the previous phases: the centering of focused attention, absolute presence of sustained concentration, silence, and absence of definition and judgment of deep contemplation. But if it's true that the whole is more than the sum of its parts, then this state will contain an additional secret ingredient: awareness. Awareness of what? It's a state of pure consciousness of being, without names, definitions, or limits. It's simply an awareness of being. In this condition of full existential potential, devoid of forms or attributes, only the pure awareness of being shines in the mind, which brings with it lightness, happiness, and joy for the simple fact of existing. It's a natural condition of being that is normally clouded by an infinity of mental activities.

In a review published in 2000 in the *International Journal of Psychotherapy*, Alberto Perez-de-Albeniz and Jeremy Holmes identified the following components common to all meditative methods:

- Relaxation
- Concentration
- Altered state of consciousness
- Suspension of logical and rational thinking processes
- Aptitude for self-awareness and self-observation

MEDITATION IN CATHOLICISM AND ISLAM

Catholicism intends meditation as a form of interior prayer closely linked to the thought of God and reflection on his word. It often begins with the invocation of the Holy Spirit so as to bring light, and continues with the reading or contemplation of an episode of the holy scriptures, analyzing its meaning, reflecting on a key word or concept, and asking God for the grace to experience the mystery contemplated. It's a commitment to make a concrete gesture to transform the message received into a charitable action and ends with giving thanks to the Lord.

In Islam, the concept of meditation is expressed by the Arabic words *tafakkur* and *dhikr*. In the mystical practice of Islam, dhikr Allah, the invocation of the name of God, is used to reach the state of meditation. Dhikr as a spiritual method of concentration was developed by the mystical movement of Sufis. This practice, developed in the Islamic world in the ninth and tenth centuries, consists of repeating one of the ninety-nine names of Allah or sacred words under the direction of a master, called in Arabic *shaykh* or *murshid*, meaning "guide."

WHY IS IT SO IMPORTANT TO LEARN TO MEDITATE REGULARLY?

The first reason concerns mental and physical health. The effects of meditation practice have a profound impact on three main levels: biological, mental, and emotional. Many scientific studies, which will be reviewed in the next chapter, show that meditating regularly, even for short periods (three months of constant practice), acts on DNA by slowing down aging and reducing inflammation.

Another great benefit is mood improvement along with the treatment of anxiety and depression. In a study conducted at Johns Hopkins University in Baltimore, after sifting through nearly nineteen thousand meditation programs and selecting the forty-seven best-designed ones involving over thirty-five hundred people, the researchers concluded that meditation can help reduce pain as well as improve depression and anxiety symptoms, and stated that there is strong evidence to recommend meditation clinically, either as a primary treatment or add-on treatment for people suffering from anxiety, depression, or chronic pain. In addition to lowering the risk of depression, meditating has a positive effect on brain chemistry as it slows the release of cytokines, an inflammatory chemical that alters mood and over time can lead to depression.

There is also important research focused on concentration, the ability to work under stress, and memory. An interesting study conducted by researchers at the University of North Carolina showed that even just twenty minutes a day of meditative practice allowed students to improve their performance on cognitive ability tests, in some cases with results even ten times better than nonmeditating groups. Similar results were observed when the participants performed tasks that required information processing and were designed to induce deadline-related stress. The experiment showed that meditators had a thicker prefrontal cortex, and the researchers concluded that meditation could compensate for the loss of cognitive abilities that occurs with old age. Long-term meditation practice also increases gray matter density in areas of the brain associated with learning, memory, self-awareness, compassion, and introspection. Finally, according to the American Psychological Association, subjects who practice even just a few minutes of meditation a day show a greater sense of social connection and positivity toward others. These results suggest that this easy-to-implement technique can help increase positive social emotions and decrease social isolation.

The extraordinary achievements of modern science essentially confirm the experience of millions of practitioners: that meditation keeps mind and body healthy, helps prevent disease and develop more meaningful relationships, and improves performance in virtually any activity.

The second reason why you should learn to meditate regularly relates to quality of life. We are immersed in a culture that celebrates knowing how to do, how to have, and how to appear as fundamental values of life. Success, happiness, and well-being are inextricably linked to these three aspects. The result is an endless competition for achieving ever more ambitious goals at the cost of everything and everyone. Most people sacrifice health, well-being, and quality of life to these three modern gods: doing, having, and appearing.

In a consumer and competitive approach to life, the trend is about things being "disposable": disposable food, disposable emotions, and disposable relationships. The state of constant euphoria, animal instinct, or physical and mental numbness are effective anesthetics.

The antidote is to rebalance yourself by rediscovering "knowing how to be." And in order to know how to be, you need to get rid of all the superstructures that generate pressure and stress. You need to undress and free yourself from cover-ups, false needs that are not yours, relational burdens, pains, and wounds. Naked like when we were born. And pure. Present and awake in the miracle of life.

Meditation is a way to counter the vortex caused by the compulsive making-having-appearing pattern, disconnected from the expression of authentic and high values. Meditation slows down and stops this vortex, leading us to really listen to what primarily exists within ourselves.

Meditation leads us to recognize, celebrate, and awaken the miracle of life in us, perceiving it in its original essence. While for most people relationships are just plugs and anesthetics for their insecurities, fears, loneliness, and frustrations, in meditative listening all this mental and emotional whirlwind ceases. All the unaware impulses that inhabit our mind and dominate us, preventing us from feeling what we really need, gradually fade away. The listening produced in meditation leads to a sense of completeness, wholeness, and happiness in oneself for the simple fact of existing. Recognize yourself in the beauty, wonder, and privilege of existing, in this form, place, and time. Now. Just exist. Meditation, if practiced and understood in depth, leads us to understand the importance of knowing how to be, even before knowing how to appear, do, and have.

Slow down, listen to each other, recognize each other, and follow each other, and realize yourself. Meditation gives us all of these gifts. When one transforms one's life into a conscious act, everything is meditation. Life itself becomes meditation. Practice, on the other hand, is the punctual act, a ritual of celebration, like a cake with a candle. You celebrate your birthday at least once a day. The choice is renewed every day. It's the soil in which we plant and cultivate seeds, virtues, and values.

The third reason why you should learn to meditate regularly regards new forms of knowledge, more evolved and refined than the common ones. Meditative practice and meditation as a state of consciousness develop, through consistency, perseverance, and patience, a particular type of knowledge called "knowledge by identification" or "cognitive absorption" (the word in Sanskrit is *samadhi*). The mind takes on the characteristics and form of what one wants to know, becoming it in perceptive and cognitive terms. Knowledge by identification doesn't simply mean imagining becoming what the mind observes but instead is something much more intimate and refined; in cognitive absorption, meditation (dhyana) is so mature and deep that it makes the impression of being separate from things and making them totally disappear. The distinction between the perceiving subject, perceived object, and their relationship dissolves. "Being what you want to know" is the highest form of knowledge and at the same time a fundamental evolutionary necessity for human beings to have a chance to survive on this planet.

The ancient Indo-Vedic wise traditions had thoroughly understood the nature of the mind, so much so that they deeply explored and completely mastered this particular form of knowledge by identification, and enjoyed all the benefits in terms of health, quality of life, and evolution. What is knowledge by identification? Is it an experience accessible to everyone, or do you need to meditate for many years and have specific qualities and characteristics? Is there a need for a radical life choice, or can anyone, with little effort, get to experience it?

The good news is that this form of knowledge is natural, simple, and within everyone's reach. What is the most exciting film you have seen at a cinema? At least once, everyone has watched a film that deeply engrossed them—so profoundly that we forget that we have a body, that we are sitting in a movie theater watching a film, that we are an entity separate from what we are watching. This total identification happens little by little: we slowly sink until we "become" the film we're watching, which comes to life through us. Attention plays a decisive role in this type of knowledge by identification. We live immersed in what is happening in the movie as if it was real. We have completely forgotten our body, the actual ego, our personality, and the place where we are, and we rule out from the cognitive and perceptive experience anything other than what happens in the film, which has become immersive, all-encompassing, and more real. We feel in ourselves what the protagonists feel. History belongs to us enough to enter into it. When contemplation also becomes intense and sustained, you enter a part of the meditative state of consciousness that is indicated by the term samadhi, literally the union between the perceiver and the thing perceived. You become what you want to know. It's not about imagination but rather about an intimate identification by the nature of our mind and use of attention. For this to occur, however, we must be deeply involved with and attracted by the object of attention, so much so that we are completely "absorbed" in it.

The same happens when reading a book. We get to a point of engagement where we fully enter the story. These perceptive and cognitive phenomena are absolutely normal, and an infinitely more satisfying, fulfilling, and profound way of knowing than what we usually use. Even two especially passionate lovers will have noticed that in their ultimate union, the perception between the lover and loved one is canceled in order to leave room only for love.

Meditation, when practiced earnestly under good guidance, purifies the mind and brings it into a state of perfection through the repeated experience of samadhi. If these forms of knowledge developed through meditation were adopted by the public educational system, it would drastically reduce drug addiction along with many problems related to health, quality of life, and relationships.

Human beings, dissatisfied with their cognitive and perceptive experiences, live in a constant state of frustration that they try to balance through surrogates and external expedients, such as drugs, addictive relationships, and distorted sexuality. Satisfaction and balance can never be achieved completely through the path of doing-having-appearing if awareness work is not first carried out in the sphere of knowing how to be.

THE SAMADHI

Technically there are two main types of samadhi, related to two particular states of mind. The first is called *samprajnata*, and it's achieved when the mind is in the state of ekagra, and involves a process of attention and concentration totally focused on a certain object (concrete or abstract), excluding any other source of attention, thought, and idea. Samprajnata is composed of *jna*, which means "knowledge," and *pra*, "superior," indicating a higher knowledge than from the ordinary levels of perception. The mind is emptied of thoughts, ideas, and impressions, regenerating itself and returning to a condition of simplicity and essentiality. The resulting clarity activates the sattvic forces, leading us to intense experiences of awareness and existential happiness.

In science, this type of meditative process in which one is asked to voluntarily focus one's attention on a chosen element (for example, breathing, or a sound, idea, or body part) has been named focused attention meditation (FAM). It has been shown that in *closed-skills* sports such as archery, golf, and gymnastics, the environment of action remains relatively stable and reliable over time, and therefore attention needs to be sustained on a predetermined sequence of actions; for this reason, the practitioner benefits more from a FAM type of meditation. Research carried out on golfers and shooters has found that meditation can reduce precompetitive stress in just four weeks, lowering cortisol (stress hormone) while improving the athletes' shooting accuracy.

The second type is called *asamprajnata*; it concerns the state of niruddha, and implies that all mental activity is inhibited and ceases completely; there is no object of attention, thought, idea, reflection, reasoning, or impression. There remains only a state of pure presence and awareness of being, beyond all definitions, forms, and identifications. This type of meditative process, called open monitoring meditation (OMM), has been studied by science and found to be and useful for evaluating experiences from moment to moment, without having to concentrate on a particular situation. Studies showed that in all *open-skills* sports, such as team sports, combat sports, and sports in which the external environment is constantly changing, one benefits from a meditative training aimed at the OMM mode.

The fourth reason, the most important, concerns the evolution and survival of the human being on this planet. Deep and authentic meditative states have side effects on health, well-being, and quality of life. Yet these aspects are just secondary manifestations of ever-higher states of consciousness as well as ever-higher awareness of oneself and life. When the mind truly immerses itself in a meditative state, it can understand life in a completely different way than ordinary and more superficial states of consciousness. The most correct term to describe what happens is *intimacy*: from this type of knowledge, a feeling of innate intimacy with the whole of creation arises, thereby allowing us to recognize the essential nature common to everything around us. The result is a deep awareness of interconnection, reciprocity, and interdependence, which generates empathy, respect, and a new sense of belonging. We become intimate with life and ourselves as never before.

The experience of knowledge by identification cancels the sense of constant separation that human beings experience as a default in their perceptive and cognitive systems. This perceptual fracture feeds the sense of isolation and separation on which we have built our sense of identity. We feel separated from others, from life, from the things that happen to us and that we see. We have lost touch with the essence common to all creation. But we can experience it again through "being what one wants to know."

Meditation, when practiced correctly, gives us new eyes to see life as we have never been able to see and understand it.

So why is meditation so important in transforming society and communities? Because it brings us together with the essence that we recognize in everything: the more we go inside, the more our consciousness expands, overcoming the sense of separation, and including others and nature. It is necessary to carry out profound work on oneself; when we are no longer strangers to ourselves, even the stranger we seem to see in the outside world is no longer a stranger. Hence the importance for a new sense of justice, community, education, and inclusion.

Our destiny is forged from the intimacy of ourselves and our level of awareness. And it can become a choice.

14   THE SCIENCE OF MEDITATION

IMMACULATA DE VIVO

Nothing moves you forward faster than correct, scientific meditation and powerful, altruistic service to the world.

—Benjamin Creme, *The Ageless Wisdom*

The interest of science in meditation has been strongly increasing in recent
years due to encouraging results. In 2019, a team led by Elissa Epel of the
University of California at San Francisco published a review comparing
nineteen different studies on the correlation between meditation and telomere length. Observing groups of different people and various meditation
techniques, these studies identified the protective effect of these practices
as they influence the mechanisms of both acute and chronic stress. We
know that stress is one of the most active factors in accelerating telomere
attrition through the action of cortisol and other hormones connected to
the fight-or-flight response. Studies show that meditation can influence
the processes that underlie the acute stress response, lowering its intensity.
Meditation is also a tool to mitigate the severity of the stress response when
it happens. Meditation antistress action is associated with longer telomeres
in subjects who practice compared to those who don't.

MEDITATION, TELOMERES, AND CHRONIC DISEASE

Additional evidence comes from a study conducted by my laboratory at
Harvard Medical School and focused on the health effects linked to a particular practice: loving-kindness meditation (LKM). This practice, derived
from a Buddhist tradition, is based on the concept of unselfish kindness

and open attitude toward others. Our study consisted of long-term meditators, defined as practicing daily sessions for at least four years and taking
part in various retreats lasting at least three days. We observed that the
practitioners' telomeres were longer than the control group that had never
done meditation, thus supporting the buffering action of LKM in maintaining DNA stability.

It's interesting to note that the benefits of meditation have been seen
not only in expert subjects who practice it regularly but among beginners
too. In 2019, a study done at various US universities observed the processes
of cellular aging in a population of 142 people aged between thirty-five and
sixty-four who practiced different types of meditation for the first time. The
practice lasted only six weeks, but it was enough to allow scientists to notice
a difference among the various types of meditation. This study found that
LKM was in fact linked to stronger telomere protection, consistent with our
results at Harvard, while mindfulness meditation was less effective.

Scholars have had particular interest in analyzing the health conditions of subjects who do meditation retreats, "full-immersive" experiences
with significant effects on telomeres. In 2018, a study conducted at several
California universities was published, correlating the impact of intensive
meditation experiences and telomere length. People in the observed group
participated in a one-month retreat dedicated to insight meditation, and
their DNA was compared with that of a control group made up of non-
meditators of a similar age, health condition, and lifestyle. Comparing
their DNA, scientists noticed a significant difference in telomere length,
suggesting that meditation retreats can be a powerful tool to promote cellular longevity.

In general, practicing some form of meditation for even short periods
of time has been associated with markers of physical and mental health.
A study done at Southern Illinois University sampling pharmacy students
observed the effects of a ten-minute daily meditation routine through a
dedicated app on various indicators of mental well-being. It was found
that the subjects who followed this program, which lasted four weeks,
showed greater inner awareness and lower levels of perceived stress. The
effectiveness of mindfulness meditation on the observed population was
uniform and significant because the participants were subjects immersed in
a demanding academic setting, therefore constantly pressured and stressed.

Similar results were observed in another 2009 study conducted at several research institutes in Atlanta. The researchers analyzed a particular type of meditation, one that promotes compassion, in an attempt to understand how it affects the stress response. The sixty-one healthy adults were divided into two groups; one followed a course of compassion meditation for six weeks, and the other was recruited as a control group. After the training, both groups were put into stressful situations, and the researchers found that those who had followed the meditation program responded better to negative stimuli. This suggests that compassion meditation can buffer reactivity to stress.

Preliminary studies are underway to assess the correlation between meditation and chronic disease prevention, such as for neurological diseases. Like a bomb squad that defuses a bomb ready to explode, meditation does so by lowering stress and interfering with the biochemical mechanism of acute or chronic response. This type of mechanism is also a factor in the development of Alzheimer's, for example. This neurodegenerative disease is characterized by the progressive deterioration of brain connections that causes significant loss of memory, especially short-term memory. Patients find it difficult to formulate words and sentences, understand verbal messages, process information such as where they are, and identify even the closest, dearest people. They struggle to name and remember the function of commonly used objects to do simple daily tasks like getting dressed. It undoubtedly has a genetic component, but environmental factors can impact its onset and development. These nongenetic components can therefore be modified for disease prevention.

A 2015 review article published in the *Journal of Alzheimer's Disease* by US researcher Dharma Singh Khalsa compared the results of various studies, particularly focusing on the effects of kirtan kriya, a simple and easily accessible meditation technique that takes just twelve minutes per day. The results are encouraging; the practice has been successfully used to increase memory in people with cognitive decline or who are severely stressed, and hence considered at risk of Alzheimer's. It has been found to be useful for improving sleep, reducing depression and anxiety, lowering inflammation, and increasing the immune response as well as normalizing insulin response. From the biological to the psychological viewpoint, kirtan kriya has been associated with an improvement in inner well-being, an important factor for maintaining cognitive functions and preventing Alzheimer's.

The beneficial effect of this meditation technique seems to be enhanced when included in a program with physical exercise, mental stimulation, socialization, and sound dietary practices.

Many studies have shown the effects of meditation in the treatment of other various illnesses, such as irritable bowel syndrome, psoriasis, fibromyalgia, anxiety, PTSD, and depression. Over the years, there has been growing interest among scientists to understand the power of meditation and awareness practices, and how this interacts with health and disease. In the mid-1990s, there was only one study available on this topic, ten years later 11 had been published, and in the years between 2013 and 2015, that number had suddenly spiked to 216 studies.

MEDITATION AND THE BRAIN

Neuroscientists in particular have tried to understand what mechanisms meditation triggers in our brain, observing which areas are involved and the type of responses. Gaëlle Desbordes, a Harvard neuroscientist, began studying the effects of meditation on the brain from her personal experience. Desbordes approached meditation as a tool to relieve the stress of professional and academic life. She experienced profound benefits from this new inner journey and decided to deepen her studies on the subject as a scholar. In order to offer these techniques to the public as a form of therapy, it's necessary to scientifically understand the benefits. A fundamental tool that she uses for her research is fMRI. This instrument doesn't just "photograph" the brain but records activities during the scan too, offering real-time reactivity to the stimuli provided. Desbordes observed the brain activity of several subjects for two months as they went about their daily routine and when meditating. Comparing fMRI results, Desbordes saw that the activation patterns of various brain areas had changed over time, especially in the amygdala, the region that regulates emotions. Initially, when participants looked at images with strong emotional content, the amygdala was intensely stimulated; by the end of the observation, Desbordes noticed that its activity was attenuated, which is likely an indication that meditation could mitigate emotional stress. What is most striking about Desbordes's studies is that the changes induced in the brain by meditation are stable, and can be detected even when the person is not meditating and instead engaged in other activities. Encouraged by the significant results of this 2012 study, Desbordes decided to go on, focusing on the effects of meditation in treating depression. These and other studies have shown that mindfulness-based interventions are effective in alleviating depressive symptoms and reducing relapse. Similar results have been demonstrated in the treatment of other psychiatric conditions such as anxiety, bipolar illnesses, eating challenges, and substance abuse.

The effects of meditation on brain structures was also studied by Sara Lazar, a Harvard neuroscientist and one of the first scholars to investigate this type of correlation. Just like Desbordes, Lazar was interested in the scientific underpinnings of mindfulness practices from her own personal experience. Her approach to meditation was completely accidental: during a training session for the Boston Marathon, she got injured, and her physiotherapist suggested she suspend training for a while and just stretch. Lazar therefore joined a yoga class, which she attended solely as a form of physical activity. During the sessions, her instructor explained that yoga would allow them to increase their sense of compassion and open their hearts to others. Lazar was skeptical about and uninterested in these aspects. Over time, however, she began to notice that she felt calmer, more focused, and able to better handle difficult situations. Her empathy toward others had actually increased, and she started seeing things from a new perspective. Just at that time she finished her PhD in molecular biology, because of her recent living experience with yoga, decided to focus her postdoc studies on the interactions between meditation and the brain. Today, Lazar teaches at Harvard Medical School and does research at Massachusetts General Hospital in Boston, where she continues to conduct groundbreaking studies.

The early studies focused on brain structures in long-term meditators, compared with a control group. Meditators had more dense gray matter in the insula lobe—an area that regulates awareness of one's inner physical states—and the areas of sensory perception, especially hearing. That's because meditation "trains" the brain to perceive the body and self, and expands sensory abilities when one focuses on the breath as well as sounds and experiences the present moment. More gray matter was also seen in the frontal cortex, linked to functions such as memory and decision-making. We know that the cortex shrinks with aging, impairing the ability to solve problems or remember. Analyzing this region, fifty-year-old meditators showed the same amount of gray matter that we find in twenty-five-year-old people. At this point the question was whether these subjects had greater amounts of gray matter even before starting meditating. A new study was therefore launched to compare brain images of a group of beginners, observed before and after a meditation course based on stress reduction through mindfulness. After just eight weeks, thickening was noted in four different areas of the brain: the posterior cingulate cortex, involved in mind-wandering activities as well as managing attention and the perception of oneself; the left hippocampus, which regulates learning, cognition, memory, and emotional response; the temporoparietal junction, associated with the ability to put things in perspective as well as compassion and empathy; and the pons Varolii, an area that produces a large amount of neurotransmitters. Conversely, the amygdala activity, which regulates the fight-or-flight response, and hence stress and anxiety reactions, was reduced. Meditation, or more generally awareness exercises, practiced even for a short period of time, can shape some areas of the brain. It strengthens those areas that govern self-perception and positive emotions along with attitudes toward others, and inhibits those that trigger negative stress responses.

In light of her many years of research in this field, Lazar recommends considering meditation as a workout that sparks something similar in the brain to what physical exercise induces in the body. It does this not only by modifying the targeted organ—the brain—but by bringing general health benefits too, helping us to better manage stress and promoting longevity. Like exercise, meditation cannot cure everything, but it can certainly be used as a supplement to other types of therapy, both psychological and pharmacological. Many health care institutions are adopting meditation as an aid to treatment. Just to give an example, at the Dana-Farber Cancer Institute in Boston, meditation courses have long been available to people undergoing cancer treatment so as to improve patients' emotional state and promote the effectiveness of treatments. It won't be the optimal weapon against cancer, but it's an additional tool that can help. As often happens, it has generally led to good results, though not for everyone. There is an individual component that can condition the outcome of these interventions or attenuate their effectiveness; it is inevitable. Even pharmacological treatments sometimes encounter limitations, but unlike drugs, mindfulness exercises have no side effects. Attempting to improve the quality of life through these practices can be a reasonable lifestyle choice, even in the absence of disease, against the anxieties and bad moods of everyday life. You don't need to spend an hour a day meditating to get some benefit. Studies so far have shown that even practicing meditation for just a few minutes a day will give good results.

This effect was highlighted by a 2017 study conducted at the University of Seoul in South Korea focused on a particular technique known as gratitude meditation. While monitoring heart rate and brain activity, scientists observed that just five minutes a day of gratitude-based meditation for a month helped subjects better manage stress, control emotions, promote motivation, and increase the level of satisfaction with their life through a mechanism of lowering the heart rate while activating the areas responsible for positive emotions.

The beneficial effects of awareness practices are long term because they modify brain plasticity. They do this by not only increasing the gray matter in the affected areas but also making the network of neuronal connections more dense. A 2012 study conducted by the Department of Psychology at Emory University in Atlanta showed that subjects more experienced in meditation had greater connectivity in attention-related neuronal networks. These areas are directly involved in the development of cognitive skills that allow one to maintain attention and not get distracted. This "altered" connectivity is found when not meditating too—a sign that this type of cognitive ability remains and is transferred to everyday life.

Type of Meditation	Effects on Health
Loving-kindness meditation	Relieves anger, frustration, resentment, and interpersonal conflicts. Protects telomeres, reduces depression, anxiety, and stress, and relieves PTSD symptoms.
Mindfulness meditation	Reduces rumination and negative emotions, improves relationship satisfaction, and attenuates impulsive emotional responses.
	Improves concentration and memory, lowers blood pressure, reduces stress, and relieves symptoms of bipolar illnesses, depression, anxiety, and eating challenges.
Breath awareness	Promotes emotional control and reduces anxiety. Improve concentration and memory.
Yoga kundalini	Increases physical strength, reduces pain, and promotes mental health.
Zen meditation	It has similar effects to mindfulness meditation, but requires more practice.
Gratitude meditation	Increases general well-being, fights depression, improves relationships with others, and helps regulate sleep.

One of the main promoters of mindfulness in the West is Jon Kabat-Zinn, a US biologist who in 1979 invented a stress reduction program based on meditation that has been adopted over time by private and public companies and institutions all over the world. In the meantime, many new techniques have been developed and general acceptance toward meditation has increased, supported by the wealth of scientific studies that confirms its beneficial effects. Mindfulness and relaxation classes are offered in a myriad of different settings, from corporations to sports teams, schools, universities, prisons, and hospitals. Even the US Army adopted a program to increase the ability of the military to manage job-related stress through meditation. Science has accumulated evidence on the benefits of meditation on physical and mental health, both when meditators are healthy—increasing well-being and quality of life—and when they're sick—with meditation lowering stress and improving emotional management—thus offering an important support to therapies and treatments.

15 MUSIC AND SOUND: HEALTH, WELL-BEING, AND LONGEVITY

DANIEL LUMERA, IN COLLABORATION WITH EMILIANO TOSO

> If the world were to end tomorrow, what would you do?
> I would leave the music on.
> —Anonymous

STOP AND LISTEN TO THE SOUND OF LIFE

Washington, DC, rush hour, subway station on a cold January morning. Thousands of people pass by. A man starts playing the violin. For about forty-five minutes, he plays six pieces by Johann Sebastian Bach. Someone stops and listens for a few moments to the melody that comes out of his instrument. After a few minutes, a woman drops a dollar in the box on the ground as she rushes by. Someone listens for a fleeting moment and then walks away; they don't want to disrupt their day. In those forty-five minutes, only six people stopped to actually listen. Children, most interested in listening, are pulled away by their parents in a hurry. The musician gets $32 total, from about twenty passersby. At the end of the performance, they collect the money, put their instrument away, and leave. No applause. No acknowledgment.

This was a famous social experiment set up by the *Washington Post* to observe people's perception, taste, and priorities. The violinist, incognito, was Joshua Bell, one of the greatest musicians in the world. On a $3.5 million violin, he played some of the most complex and beautiful pieces ever written in the history of music. Just two days before, one of his concerts

---

Emiliano Toso, PhD, is a cell biologist and composer at 432 Hz.

at Boston's Symphony Hall had sold out. Those who had the privilege of listening to him on that occasion had to spend about $100 for a seat in the orchestra. This story leads to many questions about our ability to recognize talent and beauty even in an unexpected and decontextualized situation. How was it possible that among those thousands of people, hardly anyone found a minute to stop and listen to the best music ever written, performed by one of the greatest musicians alive? Is life really something that happens while we are busy doing something else? We miss so many beautiful things if we are not able to listen to the miracle of life as it flows, if we're unable to stop and know how to listen. First of all, know how to listen to yourself. Just now. At this very moment.

That's what this is about. What are we missing right now? The symphony of life is happening right now. The show is onstage. And it's a one-off.

STOP AND LISTEN

- To the sound of nature in the forest—for fifteen minutes in silence
- To the sound of falling rain—for fifteen minutes in silence
- To the birds chirping in a tree—for fifteen minutes in silence
- To the sound of a flowing river—for fifteen minutes in silence
- To the sound of the undertow—for fifteen minutes in silence
- To the sound of the waves—for fifteen minutes in silence
- To the sound of the wind—for fifteen minutes in silence
- To the sound of silence—for fifteen minutes

THE FIRST SOUND

How many people can claim to remember the first sound they heard? The beating of our tiny heart starting in our mother's womb: that was the first mantra we heard. We listened to it for about nine months, in tune with our mother's heartbeat. The most recent studies in the field of neurophysiology and molecular biology show that the sound environment in which we are immersed from intrauterine life and then throughout our existence has the power to affect health, longevity, and self-healing processes. It does so by influencing hormones, enzymes, biomarkers, emotions, and mental state, thus affecting our whole psychophysical balance. We can perceive sound not only through hearing but also through the skin and spread it to all cells through the liquid element. Our organism is formed in the primordial water of the amniotic fluid and retains this element as its main component throughout existence; organic liquids function as a resonance instrument for the sound waves that propagate in our body.

Ludwig van Beethoven was able to express his extraordinary musical talent even after losing his hearing because he was able to hear and recognize sounds through the body.

MEDITATING ON THE SOUND AND RHYTHM OF THE HEARTBEAT: PAIR LISTENING EXPERIENCE

Place your ear on the other person's chest, at heart level. While listening to their heartbeat, hug the person. Recognize in that heartbeat the first sounds your ears heard: your little heart and your mother's heartbeat. Listen to that sound, that rhythm, and focus on what it evokes in you for five minutes. At the end of the practice, look the person in the eye and thank each other, and then swap roles.

THE MOZART EFFECT

The Mozart effect is the popular name of a study titled "Music and Spatial Task Performance" published by physicists Frances Rauscher, Gordon Shaw, and Catherine Ky in the prestigious scientific journal *Nature* in 1993. According to the research, listening to the *Sonata in D Major for Two Pianos (K. 448)* by Wolfgang Amadeus Mozart would induce a temporary increase in cognitive abilities. A group of thirty-six students underwent an abstract spatial reasoning test after experiencing one of three listening situations: Mozart's *Sonata in D Major*, verbal relaxation instructions, or silence. After listening to Mozart, the results found a temporary improvement of up to fifteen minutes in spatial reasoning (measured on the spatial reasoning subtasks of the Stanford-Binet IQ test). The Mozart effect is limited to spatiotemporal tasks involving mental imagery and temporal ordering.

EFFECTS OF SOUND AND MUSIC ON THE CELLS

There are artifacts of musical instruments dating back forty thousand years. Music arises from the need to participate in community, from the need for socialization, in order to understand the interrelationship of the phenomena of the universe and the place of humans within it, but also to ritually mark the different events of life. Music gives birth to thoughts and emotions too high and intense to be expressed by ordinary rudimentary languages. Today, science demonstrates how much music can improve human well-being by conveying not only emotions but also cellular information. Its power derives precisely from the fact that—especially if created with certain characteristics (structure, tuning, and intention) and listened to at a low volume—it acts on the human being immediately and on multiple levels—physical, biological, emotional, cognitive, and spiritual.

So what are the effects of music on our body?

CLINICAL LEVEL

Music can be used for physical and physiological balancing due to the effects it has on heart rate, respiratory rate, perspiration, body temperature, skin conductance, muscle tension, and other autonomic nervous system responses. Listening to music reduces pain, anxiety, and stress. Every living organism searches for homeostasis; stress is a neurochemical response to the loss of balance and stability. The imbalance forces the organism to find ways to restore equilibrium, and therefore music is included in activities that reduce stress and are highly protective against disease.

Listening to relaxing music (slow tempo, low pitch, and no words) has been shown to reduce stress and anxiety in healthy subjects, patients receiving invasive treatments such as surgery, colonoscopy, dental surgery, and pediatric operations, and patients with heart problems. There are interesting studies on how music allows medical practitioners to reduce the doses of anesthesia at the end of surgery and support the patient in healing.

BIOCHEMICAL LEVEL

At McGill University in Montreal, scientists study how music can change the biochemistry of our body. It has been shown that listening to music affects our health in three ways: pleasure, stress, and immune function. Psychologist Daniel Levitin's laboratories are studying how music's ways are regulated by the chemical variations of hormones such as dopamine and opioids, cortisol, serotonin, and oxytocin. Music triggers neurochemical processes similar to those that are activated by food or sex. There are neuroimaging technology systems (like positron-emission tomography scans and fMRI) that highlight the areas of the brain activated by music and compare them with other pleasurable experiences.

At least once in our lives, we have all experienced how music can motivate us, making us more attentive, focused on achieving our goals. Let's think of the role of music in painting artwork, drafting a book, or any activity in which creativity, inspiration, and silence play a role. In many hospitals, surgeons and medical staff require a certain type of music to be aired in the operating room during procedures. This practice has been proven to improve attention and collaboration while lowering stress levels for health care providers and patients. There are fascinating studies that demonstrate how listening to music during a surgical operation or childbirth can act on pain perception, allowing doctors to reduce the dose of painkillers. It seems that listening to music acts on the same brain region (nucleus accumbens) as morphine. Two markers of the hypothalamic-pituitary-adrenal axis—the hormones ß-endorphin and cortisol—have been decreased by listening to relaxing music. Alternatively, listening to stimulating music increases levels of plasma cortisol, ACTH, prolactin, and norepinephrine. This demonstrates how listening to music can regulate important hormonal circuits in our body by changing the way we respond to stress, emotions, and pain.

GENETIC LEVEL

A study published by the University of Helsinki showed that listening to twenty minutes of classical music induces the expression of genes that regulate dopamine production, neurotransmission and synaptic function, learning, memory, and cognition.

These molecular biology studies show how listening to music can create profound changes in the expression of our genome. Some of these genes are also involved in learning new songs and the gift called "perfect pitch," and are similar to those that regulate singing in birds. During listening, scientists detected a decrease in the expression of other genes that cause apoptosis (cell death) or are implicated in the development of neurodegenerative diseases.

These results support previous evidence of the therapeutic value that listening to music can have in patients with Alzheimer's or Parkinson's disease.

IMMUNOLOGIC LEVEL

Studies on lifestyle have demonstrated its effects on the immune system, stress, and inflammatory markers. Good mood, positive emotions, and humor can reduce the harmful effects of stress on both primary (innate) and secondary (adaptive) immune systems. Interestingly, studies on music have shown to impact immune response too. Music can increase the activity of natural killer cells, lymphocytes of the innate immune system—responsible for defending the human body from tumors and infections. Music can have anti-inflammatory properties and regulate cytokines, which act as communicators among cells of the immune system as well as among these and different organs and tissues. But that's not all: before surgery, music was more effective than drugs in reducing anxiety. And finally, the effect of music has been proven to promote oxytocin release, the hormone of love, happiness, and contentment.

BIOPHYSICAL LEVEL

Sound, especially when produced by acoustic music, can communicate not only with hearing and soul but also with the biology of the cells. Listening to certain types of music, such as classical music, extends survival by two months in transplanted rats. Conversely, loud and unpleasant sounds increase mortality. Our body's biological responses to sound can be measured at the biochemical, molecular, and even atomic levels. The respiratory, cerebral, and cardiac rhythms are deeply connected to the emotions we experience, our thoughts, and our psychospiritual state. Many publications point to the connection between mind, emotions, and the brain and heart rate, underscoring how musical rhythm can influence the biological information that governs a wide variety of functions directly related to health and well-being.

432 HZ LANGUAGE

Historically, much attention has been paid to music frequencies. The frequency of sounds is related to the number of cycles per second of the sound wave. It is measured in hertz (Hz) and increases with the number of cycles per second. High-frequency sounds, such as police sirens, have a frequency of thousands of cycles per second. Low-frequency sounds, such as distant thunder or a bass tuba, have a frequency of a few cycles per second.

Similarly to the central dogma of molecular biology that the flow of genetic information is unidirectional, in the field of music a standard was established. Between 1936 and 1955, it was decided which intonations all the musical instruments should have. This convention represented an important step that allowed musicians from different countries to play together.

Let's imagine each frequency as a language: just as two people must adopt the same language to communicate, so the musical instruments need to be tuned to the same frequency in order to be able to communicate with each other. In the universally recognized standard musical system, the note A in the central octave—in practice, the note that is taken as a reference for tuning an instrument—corresponds to 440 Hz. Consequently, all the other notes of that instrument will have well-defined frequencies at certain intervals. This intonation was chosen after years of discussion, and the rationale seems to lie in three elements: the possibility of integrating all types of instruments that make up an orchestra; the music produced being brilliant; and constructing the instruments far easier.

Have you ever heard a piano played at 432 Hz, however? This means not only lowering the normal pitch by 8 Hz but in practice also requires several days of work by the tuner. The benefits of this type of music on our brain along with the possibility of a greater resonance with our body, cells, and planet have been supported by scientific evidence. Playing an instrument at 432 Hz generates harmonics (frequencies related to the note emitted by the instrument) that effectively resonate with the DNA double helix (the frequency of replication), maximal brain function (bihemispheric synchronization), and fundamental beat of our planet (the frequency of sound produced by the earth, called Schumann resonance). In practice, a language is produced that is much more in harmony with nature and the universe.

Intonation at 432 Hz also leads our brain waves to vibrate at a frequency of 8 Hz (lower octaves of C at 256 Hz), namely with alpha/theta waves—a more meditative state of mind.

The difference detected on a mental level between a melody played at 432 and 440 Hz is almost imperceptible. Yet with deeper listening and awareness, the second type of tuning can generate emotions of kindness, collaboration, gratitude, peace, and relaxation more easily than the brilliance and energy of the first—more captivating as well as more suitable for advertising or use by military troops. This modern return to meditation and practices related to the search for awareness and health is rediscovering this tuning.

In order to apply the great value of music in health care and education, the most important aspects are the structure of the pieces performed (rhythm, melody, and tempo), tuning (432 versus 440 Hz), and player's intention. The latter is the most difficult component to measure, but it plays a significant role in the expression of the score through the musical instrument—an aspect often overlooked in the training programs of music academies.

SOUNDS, WORDS, AND MUSIC

Music is the space between the notes.
—popularly attributed to Claude Debussy

Eknath Easwaran, a translator of Indian sacred texts and spiritual teacher who lived between Kerala and the United States in the last century (1910–1999), wrote that Sufis advise us to speak only when our words have passed through three gates. At the first gate we ask ourselves, Are these words true? If they are, we let them pass; if they're not, we send them back. At the second gate we ask ourselves, Are they necessary? If they are, we let them pass; if they're not, we send them back. At the last gate, however, we ask, Are they kind? If they are, we let them pass; if they're not, we send them back.

As important as a kind word can be, a word of hatred, envy, lies, or ignorance can be equally devastating. The importance of the word has been well known since ancient times by many of the Eastern monastic orders. Among the most common rules, we find, "Respect the silence, speak only if strictly necessary and only to pronounce words of truth," and "Talk about your companions only if they are present." These trivial indications are difficult to apply because they call us to listen and be constantly aware. Yet following them greatly increases harmony and respect in groups and among people, allowing for more effective and nonconflictual communication.

The importance of music, words, and sounds is often underestimated because we are unaware of their enormous therapeutic impact. The influence of words, and frequencies in general, on our biological, emotional, and mental status as well as states of consciousness can become powerful tools for well-being and health. Conversely, if used with malice, intent, and unknowingly, they create important psychophysical imbalances and generate situations of suffering and pain.

Music, words, and sounds can change our body's biochemistry. This is also why we should understand to use kindness in our words.

MANAS TRAYATI: THE SOUNDS THAT FREE THE MIND

The Sanskrit word *mantra* is composed of the root *man-*, which indicates the "mind" (the thought, to think) and the suffix *-tra*, which translates into "that liberates." It is therefore possible to interpret the term mantra as "thought that liberates" or "thinking that free the mind."

Mantras originate from the Vedas, from which they then spread, becoming part of cultures such as Brahmanism, Buddhism, Hinduism, and Sikhism. Mantras can be recited aloud, whispered, or even repeated only mentally and inwardly. In Sanskrit, the intonation and act of enunciating a mantra is called *ukcara*, and its ritual repetition is indicated by the term *japa*, which means "murmur"; japa mantra is hence the repetition of those particular sounds that have the function of freeing the mind and raising consciousness. Mantras allow our mind to focus totally on a single thought (abstract or concrete, personal or impersonal), freeing ourselves from all the others, and thus intimately entering into contact with its deepest and most essential characteristics. They are tools to improve health and psychophysical balance in everyday life.

Mantras are divided into two broad categories: personal and impersonal ones. The first ones refer to the personal aspect of divinity; according to this tradition, God is present in any aspect of the phenomenal world—that is, in all creation—but has also manifested themselves on earth in various forms. Every religion has identified and personalized God through a certain figure. For Christians, God manifested two thousand years ago as Jesus Christ; for Hindus, three thousand years ago in the figure of Sri Krishna; for Buddhists in the form of the Compassionate Buddha; and for Islam, through the Prophet Muhammad.

These mantras therefore refer to the personal aspect of the deity, while others celebrate its impersonal aspect: the Absolute, the Source, the Origin, the Infinite, the Ultimate Reality, the Supreme Truth, and so on. In Hinduism, this impersonal aspect is indicated by the term *Brahman*. The choice between one and the other is made starting from the characteristics of one's mind along with its inclinations toward the personal or impersonal aspect.

The use of sounds and mantras to achieve a state of self-realization is connected to the vibratory theory, which supplies an explanation for the creation of the universe, developed by Indian sages millennia ago and then recently taken up by modern physics (including the quantum branch). The theory states that the entire phenomenal world consists of vibrations. Vemu Mukunda, nuclear physicist and Indian musician, combined his scientific studies and the Indian millenary musical tradition. By combining the physical and psychic responses to sounds, *Riza Scienze* reports that he arrived at the fascinating conclusion that "every living thing is a sound."

In the Vedas, as in the Bible ("in the beginning was the Word"), sound is the primary source from which the universe began. The sound *om*, called *pranava* (from *pra*, "before," and *nava*, from *na*, "sound"), is considered the primordial vibration of the universe, present in every thing created. The ancient sages discovered that certain sounds have the power to bring us into harmony with ourselves and the cosmos, freeing the mind while offering sensations of peace and regeneration.

There are many studies that highlight the psychobiological aspects of mantras. At the La Sapienza University of Rome, a study on clinical psychophysiology, conducted by Alessandro Gelli, Michele Cavallo, and Vito Ferri, showed that the effect of mantras can change according to the type of recitation: aloud, whispered, or mental. If the mantra is pronounced softly and combined with slow abdominal breathing, it reduces myocardial workload and lowers blood pressure, both systolic and diastolic. It could thus be useful for a person who has cardiovascular problems such as angina or hypertension. Repeated aloud and associated with deep and complete breathing, it raises blood pressure and is useful for depressed people, introverts, and hypotensive people. The type of repetition and breathing, then, must vary according to the mental state of the practitioner.

EXAMPLES OF PERSONAL MANTRAS

- *Christe Eleison* (Christ have mercy, or Christ have benevolence)
- *Hare Krishna* (Fascinating Supreme Lord)
- *Om Namah Shivaya* (I surrender unto you, Supreme Lord Shiva)

> EXAMPLES OF IMPERSONAL MANTRAS
>
> - *Om* (the primordial sound, the original frequency of the Absolute)
> - *Lokah Samastha Sukhino Bhavantu* (May all beings be free and happy)
> - *So Ham* (I am That, referring to the supreme light of life)
> - *Om mani padme hum* (Oh Jewel of the Lotus, referring to the bodhisattva of compassion)

THE SOUND OF SILENCE

The master of the Kennin temple was Mokurai, Silent Thunder. He had a little protégé, a certain Toyo, a boy of barely twelve. Toyo saw that the older disciples went every morning and evening to the master's room to be instructed in Sanzen. Toyo also wanted to do Sanzen. "Wait for a while," Mokurai said. "You are too young." But the little boy insisted, and the teacher ended up agreeing.

That evening, at the right time, little Toyo showed up at the door of Mokurai's Sanzen room. He struck the gong to announce himself, made three respectful bows before entering, and then went to sit in silence before the master.

"You can hear the sound of two hands clapping against each other," Mokurai said. "Now, show me the sound of one hand."

Toyo bowed and went to his room to think about it. From his window, he could hear the music of the geishas. "Ah, I understand!" he snapped.

The next evening, when his teacher asked him to explain the sound of one hand, Toyo began to play geisha music.

"No, no," Mokurai said. "This is useless. This is not the sound of just one hand. You did not understand anything."

Toyo moved to a quiet place. He resumed meditating.

"This time I did it."

"This is not the sound of just one hand. Try again."

In vain, Toyo meditated to hear the sound of a single hand.

More than ten times, Toyo came to Mokurai with different sounds. They were all wrong. For nearly a year, he wondered what the sound of one hand might be.

Finally, little Toyo entered into true meditation and went beyond all sounds. "I couldn't come up with anything else," he explained later, "so I achieved sound without sound."

Toyo had made the sound of only one hand.

RECOMMENDED SONGS: PLAYLIST BY DANIEL LUMERA

- Armand Amar, "Poem of the Atoms"
- Jean-Guihen Queyras, Johann Sebastian Bach's Cello Suite no. 1 in G Major
- Johann Sebastian Bach, *Magnificat BWV 243*
- Ennio Morricone, "Gabriel's Oboe"
- George Harrison, "While My Guitar Gently Weeps"
- Lotte Kestner, "Halo"
- The Yuval Ron Ensemble, "Tudra"
- Deuter, "Seashell"
- Emiliano Toso, "Forgiveness"
- Sacred Sound Choir, "Mahamrityunjaya Mantra"

16 THE SCIENCE OF MUSIC

IMMACULATA DE VIVO

> Music can pierce the heart directly; it needs no mediation.
> —Oliver Sacks, *Musicophilia*

MUSIC IS MEDICINE

For the ancient Greeks, the god Phoebus, also known as Apollo, was the patron of both music and the medical arts. Perhaps, in their great wisdom, those ancient people had already observed how closely these two disciplines are linked to each other. Music influences mood, cures melancholy, and has an almost magical power over our emotions. It is perceived by the body as something similar to medicine because, indeed, in a sense, it really is.

Scientists and doctors now have strong evidence that music can improve our health. It can enhance our neuronal networks, lower blood pressure and heart rate, reduce stress hormones and inflammation, relieve pain, and help mitigate the consequences of heart attacks and strokes. And these are just a handful of biological facts that still don't capture all the positive and regenerating effects that music has on our lives. Philosophers and poets throughout time have expressed in lofty words what science has confirmed much later. Novelist Lev Nikolayevich Tolstoy, for example, would have written that "music is the shorthand of emotion"—something that psychologists also know when they suggest to free our deep feelings through music. According to tradition, the great philosopher Plato would have said, "Music is a moral law. It gives soul to the universe, wings to the mind, flight to the imagination, and charm and gaiety to life and to everything."

MUSIC FOR STRESS, ANXIETY, AND DEPRESSION

Music calms nerves and reduces stress; that's no surprise. We experience it almost every day or whenever we get the opportunity to listen to the music we like. Our firsthand experience is now supported by scientific evidence, revealing through clinical studies the underlying processes that make music so calming.

We know that cortisol is one of the hormones responsible for stress reactions. It is involved in the fight-or-flight response, which can become chronic in the long run and be harmful to our health. Scientists investigated the role of music in lowering cortisol levels and the resulting calming effect. A French Canadian study analyzed the effects of music and silence on people undergoing particularly stressful tests to understand which of the two conditions impact cortisol the most. Researchers observed a group of students undergoing various school exams and monitored the hormone concentrations in their saliva. Following a stressful event, cortisol levels stopped rising only after they listened to music; on the contrary, cortisol levels kept rising for the next thirty minutes when the students were immersed in silence. Additional evidence comes from a study conducted in Germany in 2015: people who spend time listening to music with the specific intention of relaxing get a greater benefit, and their cortisol levels go down faster.

Music therapy has been successfully used in the treatment of depression too. In a study conducted by a Finnish research team, music-based interventions were compared to the regular psychotherapy protocols. After six months of observation, patients who had followed music therapy along with standard treatments showed significant improvement in depression and anxiety symptoms—better than the control group that followed only the normal protocol. A 2009 review of five different studies found that music can also promote regular rest in people with sleeping issues.

MUSIC, A FRIEND OF THE HEART

Music is particularly effective in protecting the cardiovascular system, the most affected by stress. A 2012 report showed that people who followed music therapy treatments had a significantly lower blood pressure and heart rate compared to the control groups. The same effect was observed in the elderly population, as demonstrated by a study conducted in Hong Kong. Participants aged between sixty-three and ninety-three with similar blood pressure levels were divided into two groups; the first listened to music for twenty-five minutes a day—every day for four weeks—while the control group did not. Blood pressure was ultimately lower in the first group than in the second one.

A 2016 study at the University of Bochum in Germany tried to understand whether different musical genres can have different effects on heart health. Researchers compared the effects of listening to Mozart and Johann Strauss versus listening to ABBA on blood pressure and heart rate. It turned out that classical music had a beneficial effect on both indexes, while the music of the Swedish pop rock group had no effect. Both genres, however, have been shown to lower cortisol levels, thus suggesting that music in general has calming effects regardless of the genre. Specifically, researchers found that the best results in lowering blood pressure and heart rate were associated with the famous *Symphony No. 40 in G Minor, K. 550* by Mozart.

Music is an effective health protection strategy not just for prevention but also as a support therapy after a pathological event. A study conducted in Wisconsin followed forty-five patients who had suffered heart attacks within the previous twenty-two hours and were hospitalized in intensive care units in stable clinical conditions. Some of them listened to classical music for twenty minutes while the others continued with the regular hospital protocols. Scientists monitored the patients' functions, and observed that the group that listened to music had a decrease in heart rate, breathing, and oxygen need, but not blood pressure. The effect lasted at least one hour after listening.

MUSIC AGAINST PAIN AND OTHER AILMENTS

Scientists have concluded that music has soothing effects even against pain. A 2016 review published in South Korea examined as many as ninety-seven different studies from the mid-1990s to 2014—a large amount of data that offer robust scientific evidence. It emerged that music therapies for suffering patients can reduce the subjective perception of pain, both in acute episodes and chronic pain situations, such as those experienced by cancer patients. Cancer treatment centers are thus adopting music therapy programs as a support strategy for surgical and pharmacological treatments.

Lowering stress and anxiety, music can help patients to better endure treatments, counteracting the effect of negative emotions linked to the disease, such as anger, fear, sadness, and sometimes even a sense of guilt and embarrassment. Music enters hospital protocols in different ways, such as interactive music therapy techniques—patients and their families participate in groups playing instruments, improvising, and singing melodies—or passive techniques based on listening to recorded or live songs. Sometimes music is accompanied by art activities like drawing and painting, providing patients with different stimuli for improving their mood and well-being. This protocol is used after treatments to support rehabilitation as well.

Music has many beneficial properties for different categories of people. Studies involving older people, for example, showed that music is associated with better indicators of cardiovascular efficiency and movement coordination, which can prevent falls and accidental fractures. A study conducted in Switzerland in 2011 followed a population of 134 men and 65 women over the age of sixty-five at high risk for falling. For the first six months, one group followed an exercise program performed to the rhythm of piano music, while the remaining subjects acted as a control group. In the following six months, the second group followed the same program, while the first group remained only under observation. The music-based activity improved the participants' balance and movement skills, reducing the risk for falling, and improving stride and gait. The benefits remained stable even six months after the intervention.

To date, most of the studies have focused on the healthy effects of music listened to by individuals, usually through headphones or earphones, and not collectively. But there is an interesting, somewhat outdated, yet still significant study that analyzed the impact on mortality among people attending collective cultural events—such as concerts or theater performances. Conducted in Sweden in 1996, this study involved 12,982 subjects, and found that people who only rarely or never attended concerts or theater had a 1.57 times higher likelihood of dying during the survey than those who attended often. Casual attendees had an intermediate risk. The data were not affected by other factors such as differences in income, social network, or education. Researchers also focused on the role of music in stimulating various areas of the brain, and consequences on hormonal levels and immune system. This is a single study, and therefore taken with caution, but the size of the sample is large enough to suggest a correlation, which should be confirmed in future studies.

MUSIC AND BRAIN

All the good and pleasant things that music does for us passes through the brain, the organ that benefits the most from this positive action. There are many studies from all over the world, with high scientific reliability. But in informal discussions, I've had the opportunity and pleasure to talk about work done by one scholar in particular, Gottfried Schlaug, professor of neurology at Harvard Medical School, director of both the Music and Neuroimaging Laboratory and the Stroke Recovery Laboratory, and a dear friend and neighbor of mine. His significant studies on the power of music to transform our brains have often been the focus of our chats during dinners at home with our families, in which I got the chance to learn many aspects of this fascinating subject.

Schlaug has been involved for many years in developing music-based therapies to help stroke patients recover their brain function. Most people with this disease experience some kind of speech impairment, from slight struggles to total aphasia—the inability to use language correctly. Medicine has long noticed that people affected by this condition, even though they are no longer able to speak fluidly, can still sing or otherwise articulate sounds according to a melody pattern. This is because the area responsible for speech and language is in the left half of the brain, while the area that governs sound modulation is in the right half. Furthermore, the right hemisphere has the ability to transform its structure to compensate for any deficits on the left side. This means that when a stroke damages the area responsible for language, the right side's ability can be used to make up for the lost function. By stimulating the patient with music therapies based on the modulation of sounds, they can form words and sentences, and gradually return to expressing themselves. The results obtained by Schlaug's team have so far been surprising. One example among many is the case of a fifty-seven-year-old patient who had suffered a large lesion in the left hemisphere of the brain and was no longer able to speak. After four years of speech therapy, they had not achieved any improvement, but after seventy-five sessions of song-based music therapy, they were able to fluently say their home address and express basic needs, thus helping family members to better understand and take care of them. What's exceptional here is that these techniques don't require a specialist; they can be performed by trained relatives or friends, as long as the process is regular and intensive. Changing the structure of the brain is possible, but stroke patients are generally between the ages of fifty and ninety, an age group in which brain plasticity is less susceptible and hence requires a lot of perseverance.

Schlaug's studies on the effect of music on the brain also addressed the difference between a musician's brain compared to that of a nonmusician. Through neuronal scans, his team observed that in a musician's brain, the areas responsible for movement, listening, and spatial vision are richer in gray matter. This is because playing an instrument, especially if learned from childhood, trains a series of complex skills that involve these areas, such as translating visually perceived musical notes into movement commands, while simultaneously controlling sound output through hearing. There is certainly an innate predisposition, visible in MRIs, but there is an undeniable process of structural adaptation through exercise, which strengthens these skills.

Apparently, our brain is profoundly influenced by music—both when we learn to play an instrument and when we listen to a song or sing— even recovering speech when a disease has taken it away from us. On the other hand, there is the well-founded hypothesis that music arrived before language itself. Darwin wrote that singing evolved before speech in the human species. As with birds, song may have played a role in favoring mating. Some scientists agree, though others less so, but it is a provocative hypothesis, which makes us feel closer to what is perhaps the only human expression that everyone likes.

RECOMMENDED SONGS: PLAYLIST BY IMMACULATA DE VIVO

- Sergej Rachmaninov, Concerto no. 2 for Piano and Orchestra in C Minor no. 2, op. 18
- Queen, "Bohemian Rhapsody"
- Camille Saint-Saëns, *Le Carnaval des animaux*
- Aretha Franklin, "Respect"
- Aaron Copland, "Appalachian Spring"
- Alison Krauss, "Down to the River to Pray"

- Gustav Mahler, Symphony no. 4 in G Major
- Florence + The Machine, "Dog Days Are Over"
- Giuseppe Verdi, *Aida*
- Simon and Garfunkel, "The Sound of Silence"

17 NATURE HEALS

DANIEL LUMERA

INVISIBLE TIES

Things are united by invisible ties. You can't pick a flower without troubling a star.
—Galileo Galilei, quoted in *Koyré, Galileo e il "vecchio sogno" di Platone*, by Francesco Crapanzano

The professor of ecology came into the classroom for the first lesson with his new students. He looked at us in silence for a few seconds and then slowly said, "Insects are the structural and functional basis of most of the planet's ecosystems. If bees suddenly went extinct tomorrow morning, in less than fifty years we would face a global extinction. On the contrary, if humans suddenly died out tomorrow, in less than fifty years life on earth would flourish again." It wasn't a lesson like any other. Driven by his ego, man considers himself authorized to dispose of the planet, creatures, and resources, only following his needs and requirements. He shows no respect for other living beings or the nature that still feeds him, nor any awareness of the intimate and essential interconnection and interdependence going on among all forms of life. In the past centuries, science and religion shared an anthropocentric evolutionary vision, where man is at the center of the universe and creation. "Man" (precisely, *Homo sapiens* and male) feels he is the "master" not only of nonhuman beings and nature but also his fellow humans who are somehow "different" in terms of gender, wealth, and even culture.

WOMEN'S SUFFRAGE

In 1945, Italy was divided into the North still under German occupation and the liberated South. Legislative decree number 23 of February 2, 1945, issued by the government headed by Ivanoe Bonomi, finally granted women the right to vote. On March 10, 1946, Italian women participated in the first administrative elections, and on June 2, 1946, they voted in the referendum to choose between a monarchy and republic. In the United States, one of the countries considered more advanced and free, women suffrage dates back to 1920 (the Fourteenth Amendment), but in order for females to vote, they needed to take literacy tests and pay electoral taxes, which greatly limited the suffrage's scope. Only in 1965, with the Voting Rights Act, were the prerequisite for a minimum level of literacy and the fee repealed. The last forms of discrimination that limited universal suffrage disappeared in the United States only in the 1970s. In France, it happened in 1946, in Great Britain in 1928, in Russia in 1917, and in Germany in 1919. The year 1893 is generally considered the time in which the first state in the world, New Zealand, introduced universal suffrage, granting women the right to vote.

This is a clear example of how slow, painful, and contradictory the evolution of relationships among human beings has been over the centuries. In any case, it has been constantly marked by a social order based on a dominant rule: power relations.

This gives us the measure of human beings' level of awareness. Just look at how they have treated and still treat their own kind. Anthropocentrism has its roots in the *Homo sapiens'* belief that they are the master and center of the whole creation as well as the best species living on the planet. The anthropocentric evolutionary model is also based on the wrong interpretation of Darwin's words, according to which "the fittest survives." This phrase is intended to refer to the being who is most able to adapt to the situations (places, environments, and relationships) in which they live, but it's been long interpreted as "the strongest." This justified the rise of a competitive and nonempathetic society, where even violence has been seen as a natural and "legitimate" tool to affirm instincts and rights as well as prevail. *Homo sapiens* tends to stop "outside" the door that leads to the temple that is "inside" each person. That's the true epicenter, the place where body and spirit meet and merge. Only from here evolutions and revolutions really begin.

The paradigm "only the strongest is suitable for survival" generated and nurtured for centuries a social system that rapidly led the planet down an ever more likely self-destructive path. Scientists are almost unanimous: the anthropocentric model cannot have a future. The alternative is a biocentric evolutionary model, no longer with humans at the center of the universe, but instead life itself, in its wonderful complexity. No longer a pyramidal system with humans at the apex, but rather a vital sphere where the evolutionary forces are based on cooperation, interconnection, and interdependence. The universe and our planet in it are a single living being regulated by balanced mechanisms. Everything is connected to everything, and every change impacts the entire system.

WHAT BROUGHT US HERE

On February 26, 2016, the Intergovernmental Science-Policy Platform on Biodiversity and Ecosystem Services published their assessment report on pollinators, pollination, and food production (our food depends on pollinating insects that are under threat). Pollinating insects are bees, butterflies, ants, hoverflies, wasps, bumblebees, male mosquitoes, and beetles. The report points out the decline of pollinator insects' biodiversity, and was drafted by 124 members of governments and more than 1,000 scientists from around the world. The results speak for themselves. There are 20,077 different species of pollinating insects, and 90 percent of wildflowers depend on them. About 75 percent of food production depends on insect pollinators and is valued at $577 billion. In nature, pollinators play a vital role as a regulatory service of the ecosystem (87.5 percent of the planet's wild flowering plants, some 308,000 species, depend on animal pollination for reproduction). About 16 percent of these insects are at risk of extinction, including 40 percent of bees and butterflies; in particular, in Europe, 9.2 percent of bee species are currently threatened with extinction. About 70 percent of the pollination of all living plant species on the planet is guaranteed by tamed and wild bees, which provide about 35 percent of global food production.

It is evident that the planet and humankind are strictly dependent on these tiny insects' lives. If they weren't there, many plant species would become extinct and the costs of maintaining current levels of productivity through artificial pollination would be unsustainable. That's not all: in the last fifty years, agricultural production has increased by about 30 percent thanks to pollinating insects, while human interventions and modern farming techniques have reduced the biodiversity of cultivated varieties by 70 percent. All of these numbers underscore the disparities that have been ongoing for several decades and forecast a disturbing future. All living species are interconnected with each other to a more or less visible degree. Altering the environment and biodiversity has devastating consequences on all species and the planet as a whole. The repercussions on ecosystems in the coming years are dire.

Science, with rare exceptions, says that if there are no rapid interventions, we will soon face the sixth global extinction. The only way to avoid this is to change the current pattern of food production in order to prevent most insects from going extinct within a few decades.

Scientific data and positions are clear. A new awareness must arise, leading not just to an external change but an internal revolution as well.

WHAT IS THE ROOT OF THE PROBLEM?

The many disasters that humans have caused since they appeared on earth originate in their ego, or what we could call the "ego sapiens." The millenary traditions and cultures of wisdom (especially Indo-Vedic trend, Buddhism, and Christianity) have tried to overcome the inner perception that makes individuals, communities, and populations feel superior and separated from things, others, life, and nature. They consider themselves masters of the world; they plunder resources as predators instead of drawing on them with wisdom, respect, and love.

So how can we transform ourselves from exploiters, rapists, and abusers of other forms of life and nature into conscious custodians of their beauty and balance? The millenary wisdom traditions offer us methods that are now being validated by wider branches of modern science. These criteria explain how the human mind really works, how one can free themselves from waste and perceptive illusions, how can one purify and elevate the mind, and how important kindness, love, silence, meditation, empathy, generosity, and compassion are. These values are part of the very nature of humans (hence derives the most appropriate definition of humanity).

We can begin by acknowledging the decisive relevance of the interconnection and interdependence we experience: the seemingly invisible and imperceptible link between the internal and external environment. But is it that important to be aware of how the world we see outside has roots in the world we carry inside? "Be the change you want to see in the world" is a call from Gandhi, urging us to become aware that every change always starts within ourselves—a revolution of consciences.

Where are the past, present, and future of *Homo sapiens* leading us? Technology, the conquest of the moon, the Venus de Milo, Statue of Liberty, Brooklyn Bridge, great architectural works, discovery of the atom, Higgs boson, Leonardo da Vinci's drawings, Beethoven's *Fifth Symphony*, pyramids, inventions, theory of relativity and quantum physics, and other innumerable discoveries along with the creativity of the intellect: If we go extinct tomorrow, what would remain of all of these footprints left in human history? Nothing. Time would erase all traces of our presence on this planet.

But humans, projecting their immense ego, have built and at the same time destroyed things (not only with senseless wars), being prisoner of their own mediocrity and fragility, deluding themselves that they can sway their way through all other forms of life without consequences. Now that the damage to the planet is becoming irreversible, it's time to break down the barrier that keeps *Homo sapiens* trapped and separates what's "inside" them from what's "outside." Observing with a neutral eye and broad perspective all the wonders and horrors in human history, we can clearly see how human beings have distorted the environment and their own way of life. So Gandhi's famous call to be the change you want to see in the world comes back to the mind more urgently than ever.

FOUR MILLENARY TEACHINGS

It seems that human ego is one of the reasons why people fail to understand the value of interconnectedness. As a species, we tend to be inflexible and categorical in every field of knowledge, including science. We attribute to a single discovery the power over all of our problems. For many years we have believed that our destiny was written in our gene heritage. Investing an enormous amount of time and money, we have come to believe that we can find all the answers to our future in DNA, such as when we will get sick, from what kind of disease, and who will lose their hair and at what age. Maybe someone has even hoped to find out the date and time of their own death. DNA can be damaged by environmental factors (viruses, radiation, pollution, and mutagens), but we used to think that we had no power to change it. The widespread belief that our fate was written in our genes with no possibility of appeal was overthrown by the epigenetic revolution. We discovered that genes, millions of switches structured in specific sequences, could be turned on or off, or modified, also in response to food, exercise, and meditation. The same mechanisms apply in neuroscience. We used to believe that the brain is everything, and that mind and consciousness are just the product of a pile of neurons. But are we sure this is true? Are we really just a pile of neurons, or will science soon prove that something else exists?

The formidable human brain that we are so proud of was responsible not only for the *Mona Lisa* and Empire State Building but also for an infinite series of threats that could wipe out all life from the planet at any moment. Human-made disasters are calling into question the survival of our and other species. Great extinctions have occurred because of apocalyptic events over millions of years. Now, however, *Homo sapiens* have the potential to unleash the apocalypse. Let's reflect on it. A knife can be used to cut a cake as well as stab a person. The knife in itself is a neutral object. What makes the difference is how we decide to use it. Splitting the atom produced a great amount of energy, but millions of deaths too. Our conscience makes the difference. This is why technology will not be the solution to our survival problems. The level of awareness that uses that technology will.

This is the first teaching of millenary wisdom traditions: before anything else, try to raise your level of awareness. Harmony and survival depend on your consciousness. Everything else is a consequence, whether technology, scientific discoveries, relationship with the environment, ethics and morals, transformations in agriculture, relationships with the environment and people, politics, social transformation, health, or quality of life.

There is urgent work that we must carry out in our interior as a species. We must redefine the perception of ourselves, our role, and our importance in light of a new awareness of interconnectedness, interdependence, and evolutionary patterns. If you want to clean up the world, you have to start by tidying up your bedroom.

The impact we have on the environment and quality of life undoubtedly comes from within: it begins in the intimacy of our mind. We have seen how when we damage a single species, even an apparently insignificant one like bees, we can impact the whole environment and put at risk resources fundamental for human survival. We are profoundly and unknowingly responsible for a number of environmental, social, and mental damages.

Even a slight increase in the average annual temperature in Europe means many acres of desertification in Africa, with consequent migrations of people forced to abandon the parched areas headed toward Europe in a boomerang effect.

UBUNTU

"In Africa there is a concept known as *ubuntu*—the profound sense that we are human only through the humanity of others; that if we are to accomplish anything in this world, it will in equal measure be due to the work and achievements of others," political leader Nelson Mandela once said.

Ubuntu is the belief in a universal bond that unites all humanity. It is a rule of life, based on compassion and respect for others. Appealing to ubuntu, people are used to saying, "Umuntu ngumuntu ngabantu"—that is, "I am what I am by virtue of what we all are."

Everything is interconnected. Understanding the key factors that generate balance in complex systems depends on our level of inner awareness. We therefore extend the responsibility to our inner world and intimate experiences of every day, be it unknowingly or consciously. From this invisible inner environment we are constantly poisoning ourselves, the world, and others—either the people we believe we hate or the people we love. Imagine that you are sitting in a room, in a moment of lightness and happiness. Imagine that a person in severe depression enters that room and sits next to you in silence. After a few minutes, your mood would probably drop, even with no apparent interaction. What happened? We are so focused on the visible aspects of life that we forget the importance of what's not visible. Many choose their partner based on aesthetics and physical beauty. When they go deep into that relationship, though, they have to deal with all the invisible parts that make the difference: the emotional and mental worlds, mood, character, psychic wounds, instincts, and drives. This is the second great teaching that ancient wisdom traditions have passed on to us: in the long run, what matters most in life is not the visible aspect but instead the invisible aspect of things.

If you have a cat in your house, observe how many times a day it takes care of cleaning itself. Several. Usually after eating or in particular moments and rituals, animals clean their bodies. Both humans and animals have the instinct, transformed into a habit, to maintain constant body hygiene as a strategy for health and disease prevention. But as a "wise species," we should take care of our emotional, mental, and conscience hygiene as well. There are many tools for keeping our inner world clean and cultivating higher awareness experiences: the water of gratitude, soap of kindness, sponge of empathy and compassion, and bubble bath of love. Just as periodic fasting can be important for health and longevity, so the mental fasting of silence regenerates the mind, making it clear, limpid, and creative. As washing ourselves eliminates waste and impurities, so forgiveness allows us to free ourselves from pains, grievances, and emotional toxins in our mind. When we look at a human being, we must be aware that what really matters, even physically speaking, is hidden from sight: most of the fundamental organs are invisible to the naked eye. Let's think of radio waves, X-rays, magnetic resonance, CT scans, and the internet. The invisible can influence and govern the visible.

If you have been in Indonesia and visited Bali, you will have surely been fascinated by the terraced rice fields, a UNESCO World Heritage Site. The irrigation system used is called *subak* and has ancient origins: it was developed in the ninth century. For the Balinese, a subak doesn't simply supply water to the plants but rather is a complex and refined ecosystem. Subaks are an example of a network based on a conscious interconnection, not just an effective instance of good stewardship of the common good and shared responsibility (as well as a democratic economic system), but something much deeper.

The awareness of interconnection is expressed through the philosophical concept *Tri Hita Karana*——that is, three (*Tri*), happiness/wellness (*Hita*), and cause/origin (*Karana*). Balinese philosophy affirms that three things are the origin and cause of happiness and well-being:

1. *Parahyangan*, the harmonious relationship between the human and infinite
2. *Pawongan*, the harmonious relationship between human beings and their neighbors
3. *Palemahan*, the harmonious relationship between human beings and nature (the environment)

One of the most fascinating aspects is that the power over the common good is entrusted to the person who owns the lowest land. This is for a simple reason: if they don't ensure the correct irrigation of the upper lands, water will not reach their own land. Common good corresponds to personal good. If the leader doesn't care about the others' well-being, they won't get their well-being either. Personal responsibility puts power at the service of the common good, favoring the well-being of the entire social structure. A subak is an ancient illustration of the successful management of the common good. It relies on the awareness of some fundamental elements: interconnection, interdependence, harmony in relationships, respect for nature, and respect between humans and the cosmos. All Balinese culture is imbued with spirituality. The subak system is something sacred, connected to the meaning and purpose of all things, because human activities are intended to be in relation to the cosmos and existence. It's a way of respecting the gift of life and its miracle. All the members of the subak can cultivate rice together in harmony, even in a society strongly divided into castes.

Let's now reflect on the common element of successful companies. It's undoubtedly interconnection. Facebook is among the world's largest content producers, but it doesn't produce any content; people, interconnected through it, produce it. Amazon and eBay are the largest product sellers in the world, but they don't actually sell any products; they interconnect sellers and facilitate widespread distribution.

The third teaching of the ancient wisdom traditions is that everything has an inner root. The main problem with human beings is their boundless ego. If *Homo sapiens* really wants to evolve on this planet, they must overcome their chronic egotistical condition. The revolution of conscience is a basic step that needs to happen on a global scale. All revolutions so far have been reactions against dictatorships, abuses, injustice, and violence. They have been motivated by anger, pain, rebellion, the desire for revenge, and justice. But how much blood has been shed in the name of ideals of freedom, equality, and fraternity? Hasn't Communism turned into dictatorship? Didn't so many heads fall during the French Revolution in the name of fraternity? And weren't the Crusades also done in the name of God?

A revolution of conscience in something different. It's when one is reconciled with oneself, when inside and outside make peace. It's when you understand your responsibility along with the impact that emotions and thoughts have on everybody's destiny. It's a revolution where one renounces hatred and violence, regardless of the external conditions. It doesn't matter what happens outside; choosing love remains an option of freedom. How else could the saints have seen the love of God among the suffering people? How could Mandela have forgiven his jailers and called them to rule South Africa with him after he spent twenty-seven years in prison? By doing this, he freed himself and the whole country from hatred, rather than responding to it with more hatred. From this intimate choice, through consistent behavior, he influenced the whole world. The revolution of conscience is proactive; it doesn't react to an existing state of facts, or feed hatred, rejection, or competitiveness. It's generated by awareness. It flourishes through kindness, with determination, strength, and harmony. People who have matured in the silence of their heart have a completely new vision of the meaning of life and their role in it, and don't want to change the world; they celebrate what they have made of themselves. They share it with others as a gift.

How to get to such a radical change? Through the fourth great teaching of the ancient millenary traditions: expand your consciousness and include everything in it. Start changing your deepest sense of identity.

OVERVIEW EFFECT

"Oh my God, look at that picture over there. That is the Earth coming up," exclaimed astronaut William Anders before taking *Earthrise*, one of the most famous photographs in history. It was December 24, 1968, during Apollo 8, the first mission in moon's orbit.

Seeing earth as it is—curved, radiantly blue, infinitely small and fragile floating in a boundless black space without depth—experiencing firsthand that deep sense of unity, and losing any sense of separation between nations, countries, ethnic groups, humans, and living beings is a breathtaking experience that you will not forget.

From the earth to the moon and back in a snapshot that has forever transformed the cognitive, perceptive, and emotional experience of all astronauts, giving them and the whole world a new perspective on humanity is called the overview effect.

This effect, first theorized in 1987 by writer Frank White while observing earth from an airplane window, is receiving increasing attention from scientists, psychologists, and researchers around the world due to its profound impact. It is in fact a radical cognitive shift due to seeing our planet from a great distance for the first time and then suddenly realizing that we are all part of the same system, with no more differences or borders. Earth appears as a tiny blue dot immersed in an infinite universe, a wonderful living organism that breathes, a unitary, interconnected, and at the same time extremely fragile system.

The overview effect is about expanding one's vision and perceptions until reaching an ecstatic feeling of oneness with every living form, the planet, and entire universe, beyond one's personal history, desires, needs, and little worries, in full awareness of a global union. It is recognizing oneself no longer by one's name, role, job, social status, gender, or species but instead feeling and being present in the miracle of "being alive in this instant." Indulge yourself in the luxury of experiencing wonder at being an expression of a single life and "being together." What would it be like to make a choice from this perspective? What would change now in your life? How would you choose to live?

The good news is that you don't have to go to space to experience this shift in perspective. When astronauts and scientists searched the existing literature for traces of this effect, they discovered that it was described for the first time thousands of years earlier by ancient wisdom traditions as one of the states of consciousness that can be experienced in meditation: the *savikalpa samadhi*, a state of total fusion with the cosmic consciousness in which the illusion of separation is canceled in a dazzling sense of unity that radically transforms the vision of oneself, others, and the world. It is from this vantage point that the key to our survival unfolds. We need to start acting as one species sharing the same fate.

We won't survive if we don't.

Even science has shown how our DNA connects us to each other and life on our planet in a single family that includes plants, animals, birds, insects, fungi, and even bacteria. It's time to take on a new responsibility as tenants of this fragile and beautiful planet we call home.

A SINGLE LIVING ORGANISM

To see a World in a Grain of Sand / And a Heaven in a Wild Flower.
—William Blake, "Auguries of Innocence"

Hair in the wind, bare feet, and off we go running at breakneck speed in a meadow, among trees and flowers with that broad smile typical of children while a parent is shouting not to go too far. It's an image so familiar as to make it hard to believe that after a few years, it would have been so easy to forget the joy of nature. Yet not only are adults nowadays increasingly disconnected from nature but children also enjoy less and less of what has been their natural playground since time immemorial.

Television, technology, video games, increasingly gray and less human cities, a sedentary lifestyle, junk food, and almost everything else seems to help distance us from our nature. So much so that the intervention of science was necessary to remind us of what we all intuitively know: contact with nature heals, reduces stress, anger, and fear, is a cure for anxiety and depression, stimulates positive emotions, rekindles vitality, gives meaning to life, and reduces blood pressure, heart rate, muscle tension, and according to some research even mortality.

All it takes is a plant in hospital rooms, offices, or classrooms to significantly reduce stress and anxiety. Research conducted by Bo-Yi Yang of the Sun Yat-sen University School of Public Health in Guangzhou on 59,754 children and adolescents has shown that a small green area within five hundred meters of a school or kindergarten significantly reduces the symptoms of attention deficit and/or hyperactivity by improving attention span.

In a historical period characterized by a growing feeling of loneliness, nature still helps us; time spent in nature reconnects us to each other and the world around us. A study done at the University of Illinois suggests that those who live in areas with trees and green spaces around them tend to get to know more people, develop strong feelings of unity with neighbors, be more likely to help and support each other, and have stronger feelings than those who live in buildings away from trees and nature. In addition to this increased sense of community, the risk of street crime decreases, as does domestic violence and aggression, and the ability to cope with the demands and problems of life increases. It is enough to look at a photograph of nature for an MRI to detect the activation of areas of the brain related to empathy and love; on the contrary, when observing an urban photograph, the areas associated with fear and anxiety are activated.

But was the intervention of science really necessary to remind us of something so obvious? In recent decades we have invested exorbitant amounts of money, time, and resources in new technologies, drugs, and therapies against all forms of mental and nonmental illnesses, when next to us we have one of the most powerful, cheapest, and natural remedies ever: nature. "Our" nature. We have destroyed immense natural areas to build ever more inhumane cities and exploit the planet's resources in the name of a "well-being" that is leading millions of people to die of stress-related pathologies, forgetting that it is precisely the connection with earth that keeps us healthy. Isn't it time to stop for a moment and ask ourselves, What are we doing?

THE INVISIBLE LEGACY

Science has long been convinced that our whole life, character, and personality are the result of an unchangeable genetic determinism. Then came epigenetics, which demonstrated how we can impact our DNA, activating or deactivating specific genes thanks to nutrition, meditation, and lifestyle in general. And that's not all: recent research has even revealed that what happens in our lives is transmitted to our children and our children's children. A study at the University of Zurich showed that traumas can be handed down to the third generation. This transmission passes through microRNAs—genetic molecules regulating cellular functions—that can be altered by traumas so that these are passed on to descendants through gametes.

Similarly, studies about Holocaust survivors found that their children have the same genetic modification due to their parents' trauma, even though they never experienced any of it directly. The same thing has been observed in the children of soldiers captured by enemies during World War II; on average, they died younger than those whose fathers escaped captivity, even though they were born after the war and therefore did not experience any of that trauma. These studies reveal that the DNA of survivors—victims of atrocious suffering as well as physical and psychological torture—keeps traces of the trauma suffered and passes them on to the offspring.

This is the kind of responsibility we have to ourselves, our children, our children's children, and the world as a whole. This is something that we intuitively knew and that now science is starting to support with strong evidence. An even more significant step forward has been taken from these discoveries. For example, mice that were induced to fear red berries because they were associated with electroshock gave birth to pups who are afraid of berries even without knowing anything about electroshock. But these same pups, when placed in "positive" environmental and social settings, can defuse the potential consequences of the trauma, not only modifying their own lives, but interrupting the hereditary transmission of the trauma to their offspring.

The implications of these discoveries revolutionize the concept of transgenerational responsibility (individual, social, and collective), extending it to the inner world.

What are we leaving to our children? Are we aware of what we really leave as a legacy on this planet? Besides money, a house, and bank account, what wealth are we really passing on to our children and grandchildren? A life written with words of pain, misunderstanding, narrow-mindedness, shortcomings, and wounds? Or a life free from trauma and conditioning, happy and open to the infinite possibilities that the future can hold?

The tools to carry out a work of inner transformation are not lacking. We have one option: to choose to be brave and resolve the pending inner issues so that children and grandchildren don't pay for our ignorance.

Each of us has their own choice.

THE BUTTERFLY EFFECT

Can the flap of a butterfly's wings in Brazil set off a tornado in Texas? This was the title of a conference held by mathematician Edward Lorenz in 1972. Commonly known by the name "the butterfly effect," hence the famous movie, this phenomenon is the basis of chaos theory, according to which tiny variations in the initial conditions of a system would cause huge long-term variations in the behavior of that system.

The example proposed by Lorenz, to be precise, concerned the climatic sphere: all that is needed is the flapping of a butterfly's wings because through chain reactions (displaced molecules that collide with other molecules, which collide again with other molecules), a hurricane can be generated. This theory explains the unpredictability of weather conditions since even if the most accurate forecasts are made, the smallest variation can upset the entire system. Time, in fact, is a classic example of a chaotic system.

The implications of the butterfly effect have also recently been studied in the human brain. The researchers' idea was to cause a small "perturbation" in the brain, the neuronal equivalent of a butterfly's wing beat, and observe whether this interference would grow to affect the whole brain or instead be quickly extinguished. The outcome of the experiment showed that a small extra impulse is enough to start a chain of further impulses that will gradually affected millions of neurons. From this perspective, imagine the importance that every single thought, emotion, action, and food has on our lives. What are the impulses we are transmitting to our brain? What thoughts do we feed ourselves every day? And with what emotions? Are we feeding on fear, anxiety, greed, and anger, or kindness, compassion, and love?

Living and experiencing values such as kindness, forgiveness, peace, silence, responsibility, and empathy radically modifies the quality of life, relationships, and individual and collective well-being. Translated into biological terms, it represents a successful evolutionary strategy to survive hatred, violence, and revenge, and profoundly affects every level of our lives. When we experience a real increase in awareness, the relationship with the value of existence flourishes and a deep light awakens in the eyes that illuminates everything in our lives, giving it new meaning. Today, more than in any other historical period, it is necessary to create a bridge between science and the values of the most ancient traditions, between humans' well-being, health, and happiness, the development of our conscience, and the deep sense of interconnection that binds all beings. Sometimes a wing beat in one's inner world is enough for everything to shine with new life. This is the way of kindness.

A PLEDGE TO PLANET EARTH

DECLARATION OF INTERDEPENDENCE (DAVID SUZUKI)
United Nations' Earth Summit, Rio de Janeiro, 1992

THIS WE KNOW
We are the earth, through the plants and animals that nourish us. We are the rains and the oceans that flow through our veins. We are the breath of the forests of the land, and the plants of the sea. We are human animals, related to all other life as descendants of the firstborn cell. We share with these kin a common history, written in our genes. We share a common present, filled with uncertainty. And we share a common future, as yet untold. We humans are but one of thirty million species weaving the thin layer of life enveloping the world. The stability of communities of living things depends upon this diversity. Linked in that web, we are interconnected. . . .

(continued)

THE UNIVERSAL RESPONSIBILITY (EARTH CHARTER)

To realize these aspirations, we must decide to live with a sense of universal responsibility, identifying ourselves with the whole Earth community as well as our local communities. We are at once citizens of different nations and of one world in which the local and global are linked. Everyone shares responsibility for the present and future well-being of the human family and the larger living world. The spirit of human solidarity and kinship with all life is strengthened when we live with reverence for the mystery of being, gratitude for the gift of life, and humility regarding the human place in nature.

We urgently need a shared vision of basic values to provide an ethical foundation for the emerging world community. Therefore, together in hope we affirm the following interdependent principles for a sustainable way of life as a common standard by which the conduct of all individuals, organizations, businesses, governments, and transnational institutions is to be guided and assessed.

THE INNER COPERNICAN REVOLUTION

In 1543, astronomer Nicolaus Copernicus proposed the heliocentric theory that was supposed to replace the Ptolemaic geocentric one, opening a long philosophical dispute on the possible heresy of the scientist's work and much controversy in the Protestant Church. Copernicus's heliocentric theory was revisited by Galileo, who as everyone well knows, was investigated by the Holy Office because he affirmed the scientific validity of heliocentrism when all doubts had not yet been resolved. From the scientific level, the clash moved to the doctrinal and political one, and ended with Galileo being sentenced to life imprisonment, which was then turned into a sentence of house arrest. Galileo was sentenced to recite daily prayers for three years and also had to recite an act of abjuration in which he disavowed the "false opinion" of his theory. In Europe, thought crimes were no longer prosecuted only after World War II. Although it took until 1851, the definitive demonstration of Galileo's heliocentric theory was thanks to physicist Jean Bernard Léon Foucault (the famous "Foucault's pendulum"). In fact, after astronomer Johannes Kepler and mathematician Isaac Newton, modern astronomy has shown that the sun and Milky Way are also in motion since the entire universe is expanding.

The Copernican revolution, however, has not yet taken place within the human being, who continues to follow an anthropocentric model.

Believing that creation revolves around human beings and that our species is the most evolved form of life on the planet is a bit like still believing that the solar system and entire universe revolve around the earth. The Copernican revolution should be internalized, changing the system of beliefs and perceptions, but above all, overcoming the illusions of the ego sapiens, who tends to feel superior and separate from what surrounds them as well as from life itself.

OVERVIEW EFFECT

Let's try an experiment. We gradually expand our sense of identity, and then see what changes in the choices, decisions, and perspectives through which we observe the world and ourselves. There are multiple levels of identity. Let's imagine applying the overview effect through a simple exercise of defocusing on ourselves. Let's therefore consider a problem and observe it through different levels of identity, increasingly broad and impersonal.

Look at the problem through the perspective of your closest self-identity: your historical personality and ego. Before being a personality, your sense of identity belongs to the superior category of a man, woman, or other gender. You would realize, if you looked at the same problem from this larger order of magnitude, that some priorities, considerations, and needs could change.

I AM . . . (YOUR NAME)

How do you view the problem from this level of identity?

I AM A MAN, WOMAN, OR OTHER GENDER

How do you view the problem from this level of identity?

I AM A HUMAN BEING

How do you view the problem from this level of identity?

I AM A BEING

How do you view the problem from this level of identity?

I AM LIFE

How do you view the problem from this level of identity?

THE SEVEN SEEDS OF INTERCONNECTION

1. The higher the awareness of interconnection in the human being, the more the possibilities of harmonious evolution and survival increases.

2. There is a close connection between the internal and external environment. The intimacy of our feelings, perceptions, and mind creates reality itself.

(continued)

3. What we perceive as the internal and external world influences each other in a close interdependence. The harmony and balance of one cannot ignore the other.

4. The greatest challenge for human beings is overcoming our own ego—that is, the sense of separation that makes us believe we are not intimately interconnected.

5. Our well-being depends on others' well-being; the life of the people we love, the community, and all other beings also depends on our actions. Your and everyone's destiny depends on your level of awareness.

6. A new revolution should begin with the awareness of the impact that our intimate feelings, thoughts, and emotions have on the world. It is an inner revolution that can truly change the sense of individual and collective identity, and the role of human beings on this planet. It's a gentle revolution of awareness.

7. The value of kindness, essential and undeniable, should be the basis of any relationship among human beings, so they can relate to each other in the most beneficial, fraternal, and noble way.

18 NATURE AND WELL-BEING

IMMACULATA DE VIVO

Nature always wears the colors of the spirit.
—Ralph Waldo Emerson, *Nature*

Looking at it as a whole, the natural environment of Belmont, Massachusetts, the small town where I live, is not diverse. It's a typical New England landscape, with green fields along with a handful of characteristic trees such as pines, oaks, and maples. But when I walk in the woods near my home, I always feel like an explorer, even after so many years. Each time I notice something different, linger on a detail, and catch a variation of the light or a new smell. Looking around me, I realize the wonderful ability of the surrounding landscape to always give me new stimuli, while remaining all in all the same. It accompanies me with its curves along paths that I know by heart, but that change continuously according to the seasons, as animated by a pulsation of life and ever-changing colors. I usually walk aimlessly, without a destination. The pleasure is in the walking, and the beauty is in the journey itself. I nod to other walkers crossing my path. Any anxiety or sad feeling dissolves as I make my way through the woods. And if I walk far enough to leave everything behind, I reach a state of true joy, an inner stillness that becomes happiness. I feel relieved of all weight, satisfied to have once again escaped the oppression of the four walls, free from the bondage of sitting. Now I can move and refresh my mind by breathing clean air. A long walk is a rebirth for me. I come back different than before. I feel better, even if only for a little while.

The sense of well-being that nature can give us is an experience that I feel a lot firsthand, but everyone has experienced it. Even scientists have begun to study it so as to understand the impact it has on our physical and mental health. The results so far have been astonishing. In 2016, a study at Harvard School of Public Health by Francine Laden analyzed data on 108,630 women from the NHS, residing in the most diverse areas of the United States. Women who live in areas with green spaces within 250 meters from their home tend to have a 12 percent lower mortality rate than those who live in more urbanized areas. Mortality decreased 13 percent for cancer, 35 percent for respiratory diseases, and 41 percent for kidney diseases. These are impressive numbers.

But how does this interaction work? Obviously it's a multifactorial process. Having easy access to green spaces near home leads to spending more time outdoors, perhaps doing physical activity like walking, jogging, or cycling. You breathe clean air, and get away from pollutants and noise, and this lowers your stress while alleviating depression and anxiety symptoms. Exposure to sunlight makes your body produce vitamin D, fundamental for a healthy immune system. A vitamin D deficiency is associated with, among other things, depression. Green spaces also improve social life, favoring contact with friends and the possibility of making new acquaintants, with a huge impact on well-being. Vegetation cleans the air by lowering the levels of nitrogen dioxide and particulate matter, which irritate the respiratory tract. The mortality rate linked to lung disease among women in the study who were most in contact with nature was one-third lower than the others.

Scientists have long been suggesting that lawmakers increase both policies aimed at creating green spaces for public use and greenery in densely urbanized areas, as a contrast to pollutants and an antistress strategy.

Telomeres seem to be positively affected by contact with nature too, as shown by a 2008 study done at the University of Hong Kong. Researchers there analyzed the health status of a cohort of 976 sixty-five-year-old men residing in different areas of the Chinese metropolis, each characterized by a different level of urbanization. Health data from people who lived in four districts of Kowloon, the old part of the city, were compared with those who lived in Shatin, a recently built area with a higher standard of living. The Shatin neighborhood has buildings along the banks of a river, many parks and green areas, various shops, and a public transport terminal directly connected to the city center. The study showed that people who lived in the old city had shorter telomeres than Shatin residents, who enjoyed more contact with nature and a better quality of life.

In 2018, a collaboration among some European and US universities led to an interesting study on the benefits of nature. The objective was measuring how long one must be exposed to green spaces to stimulate positive responses in the body. The study involved a population of 19,806 people residing in England. Researchers tracked the participants' habits, health status, proximity to green areas, and time spent in contact with nature. It emerged that spending at least 120 minutes a week outdoors increases various indexes of good health. This is probably due to the environment being healthier than urbanized areas, but it can also depend on people being more likely to perform some form of physical activity, even if light.

The benefits of contact with nature for physical health were further investigated in a 2019 study, conducted by a Portuguese team in collaboration with New York University. Scientists focused their attention on the effects that greenery has on children's health, observing the health profile of 3,108 seven-year-olds from the metropolitan area of Porto. They analyzed the interactions of exposure to nature with various biomarkers related to the immune system, inflammatory reactions, and metabolic and cardiovascular systems. The researchers found that having green spaces within eight hundred meters of a school was associated with reduced stress indicators—an effect that decreased as the distance increased. The study offered a clear sign of nature's ability to "fortify" us from early years on.

Neuroscience has also investigated this phenomenon, with interesting results. A 2015 study analyzed the role of urbanization—which reduces contact with nature—on the onset of mental illnesses. Psychology has identified "rumination" (that is, obsessively brooding over a chain of thoughts that cannot be broken) as one of the risk factors for developing depression and other mental conditions. Research has shown that even just ninety minutes of walking in nature can lower rumination. The effect has been confirmed by scanning the activity of the subgenual prefrontal cortex—that is, the area involved in the mechanisms of depression. A ninety-minute walk in an urban area did not supply the same result, suggesting that nature can contribute to improving collective mental health.

If just a few hours of contact with green spaces can greatly benefit your health, what can long-term exposure do? This is the question that scientists from Spain tried to answer in a 2017 study involving 253 school-age children in Barcelona. Comparing MRI scans with data about the distance of each subject from the city's green spaces, scientists saw that prolonged contact with nature is associated with a greater amount of gray matter in different areas of the brain related to memory, attention, and other cognitive functions. Nature is the environment from which we come. It's where our intelligence was formed and our skills were refined. For millions of years we have lived among trees and flowers, and today we discover, thanks in part to science, that moving away from nature can hurt us, undermining our health and making us more fragile.

ACKNOWLEDGMENTS

Heartfelt thanks to Giada Caudullo, without whom this book would never have been written. Thanks to Anna Giuliani, Livio Gaudenzio, and the whole Giuliani Caudullo family. Thanks also to Emiliano Toso and Vincenzo Sorrenti for their contribution to the work, and Fabio De Vivo, Cristina Franchini, and Diletta Marabini for coordination and support.

Immaculata De Vivo and Daniel Lumera

To my family and friends.
    To Andrea D'Ascenzi for his precious help.
    To Alyssa Goodman, Felice Frankel, and Fran Berman for their friendship as well as our constant intellectual exchange.
    To all of my Harvard University colleagues and collaborators, who welcomed me into their world and allowed me to broaden my horizons. We work together every day to better understand nature, health, and human well-being.

Immaculata De Vivo

To Angela Cavazzuti and Alberto Pinna.
    To Felicia Cigorescu for her presence, care, and love.
    To my Master Anthony Elenjimittam.
    To all the "Golden Thread" researchers, the students of My Life Design Academy program, and all the volunteers and ambassadors of the International Kindness Movement.
    To Tara Gandhi Bhattacharjee, A. Sergi Torres, Sara Sariki, Luca Gonzatto, Mamen Díaz, Gastón Peláez, Gonzalo Rivas, Oriol López, Hanane Miftah

Ben Haki, Sol, Nur, Nuria García, Assumpta Civit, Fanny Mas Giordana, Nico Caiazza, Candida Di Bonaventura, Mauro Paolinelli, Roberto Pagani, Manuela Zavan, Serena Porciani, Valeria Pompili, Piergiorgio Baldelli, Gabriele Bardelli, Nicholas Tabanelli, Gaia Monti, Linda Ammendola, Fabio Tosoni, Yanua Ranuzzi, Simone Tani, Patrizia De Libero, Roberto Marabini, Laura Cariolato, Alessandra Celia, Rashmi V. Bhatt, Annarosa Colonna, Fiorina Bartiromo, and Giorgio Monti.

Daniel Lumera

BIBLIOGRAPHY

Adorini, L., and G. Penna. "Control of autoimmune diseases by the vitamin D endocrine system." *Nature Clinical Practice Rheumatology* 4, no. 8 (August 2008): 404–412. doi:10.1038/ncprheum0855.

Aristotle. *The Nicomachean Ethics*. Translated by David Ross. Oxford: Oxford University Press, 1984.

Aftanas, L. I., and S. A. Golocheikine. "Non-linear dynamic complexity of the human EEG during meditation." *Neuroscience Letters* 330 (July 2002): 143–146.

Aherne, C., A. P. Moran, and C. Lonsdale. "The effects of mindfulness training on athletes' flow: An initial investigation." *Sport Psychologist* 25 (2011): 177–189.

Alimujiang, A., A. Wiensch, J. Boss, et al. "Association between life purpose and mortality among US adults older than 50 years." *JAMA Network Open* 2, no. 5 (2019). doi:10.1001/jamanetworkopen.2019.4270.

Aubert, G., G. M. Baerlocher, I. Vulto, et al. "Collapse of telomere homeostasis in hematopoietic cells caused by heterozygous mutations in telomerase genes." *PLOS Genetics* 8, no. 5 (2012). doi:10.1371/journal.pgen.1002696.

Aviv, A. "Telomeres and human aging: Facts and fibs." *Science of Aging Knowledge Environment*, no. 51 (2004): pe43.

Ayesha, S. A. "Post-traumatic stress disorder, resilience and vulnerability." *Advances in Psychiatric Treatment* 13, no. 5 (2007): 369–375. doi:10.1192/apt.bp.106.003236.

Benetos, A., K. Okuda, M. Lajemi, et al. "Telomere length as an indicator of biological aging: The gender effect and relation with pulse pressure and pulse wave velocity." *Hypertension* 37, no. 2 (February 2001): 381–385.

Benson, H., and R. K. Wallace. "Fisiologia della meditazione." *Le Scienze* 45 (1972): 70–76.

Berry, L. L., T. S. Danaher, R. A. Chapman, et al. "Role of kindness in cancer care." *Journal of Oncology Practice* 13, no. 11 (November 2017): 744–750. doi:10.1200/JOP.2017.026195.

Bhasin, M. K., J. A. Dusek, B. H. Chang, et al. "Correction: Relaxation response induces temporal transcriptome changes in energy metabolism, insulin secretion and inflammatory pathways." *PLOS ONE* 12, no. 2 (2017). doi:10.1371/journal.pone.0172873.

Bilton, Nick. "The American diet: 34 gigabytes a day." *New York Times*, December 9, 2009.

Bishop, S. R., M. Lau, S. Shapiro, et al. "Mindfulness: A proposed operational definition." *Clinical Psychology: Science and Practice* 11 (2004): 230–241.

Blackburn, E. H., and E. S. Epel. "Telomeres and adversity: Too toxic to ignore." *Nature* 490 (2012): 169–171. doi:10.1038/490169a.

Blasco, M. A. "Telomeres and human disease: Ageing, cancer and beyond." *Nature Reviews Genetics* 6, no. 8 (August 2005): 611–622.

Bohn, Roger E., and James E. Short. *How Much Information? 2009 Report on American Consumers*. San Diego: University of California, Global Information Industry Center, 2009.

Bonanno, George A. "Loss, trauma, and human resilience: Have we underestimated the human capacity to thrive after extremely aversive events?" *American Psychologist* 59, no. 1 (January 2004): 20–28. doi:10.1037/0003-066X.59.1.20.

Briegel-Jones, R. M., Z. Knowles, M. R. Eubank, et al. "A preliminary investigation into the effect of yoga practice on mindfulness and flow in elite youth swimmers." *Sport Psychologist* 27 (2013): 349–359.

Brodwin, E. "What psychology actually says about the tragically social-media obsessed society in 'Black Mirror.'" *Business Insider*, October 26, 2016.

Brody, J. E. "Looking on the bright side may be good for your health." *New York Times*, January 27, 2020.

Brown, K. W., R. M. Ryan, and J. D. Creswell. "Mindfulness: Theoretical foundations and evidence for its salutary effects." *Psychological Inquiry* 18 (2007): 211–237.

Brown, S. L., D. M. Smith, R. Schulz, et al. "Caregiving behavior is associated with decreased mortality risk." *Psychological Science* 20, no. 4 (April 2009): 488–494. doi:10.1111/j.1467-9280.2009.02323.x.

Browning, R. "Balaustion's adventure." In *The Poetical Works of Robert Browning*. Vol. 6. New York: Macmillan, 1894.

Cadenas, E. "Mitochondrial free radical production and cell signaling." *Molecular Aspects of Medicine* 25, nos. 1–2 (February–April 2004): 17–26.

Calado, R. T., and N. S. Young. "Telomere diseases." *New England Journal of Medicine* 361, no. 24 (December 2009): 2353–2365. doi:10.1056/ NEJMra0903373.

"Can relationships boost longevity and well-being?" *Harvard Health Publishing*, June 1, 2017.

Carlson, L. E., Z. Ursuliak, E. Goodey, et al. "The effects of a mindfulness meditation-based stress reduction program on mood and symptoms of stress in cancer outpatients: 6-month follow-up." *Supportive Care in Cancer* 9, no. 2 (March 2001): 112–123.

Carlson, M. C., K. I. Erickson, A. F. Kramer, et al. "Evidence for neurocognitive plasticity in at-risk older adults: The Experience Corps program." *Journals of Gerontology: Series A, Biological Sciences and Medical Sciences* 64, no. 12 (2009): 1275–1282. doi:10.1093/gerona/glp117.

Carver, C. S., and M. F. Scheier. "Dispositional optimism." *Trends in Cognitive Sciences* 18, no. 6 (June 2014): 293–299. doi:10.1016/j.tics.2014.02.003.

Cassidy, A., I. De Vivo, Y. Liu, et al. "Associations between diet, lifestyle factors, and telomere length in women." *American Journal of Clinical Nutrition* 91, no. 5 (2010): 1273–1280. doi:10.3945/ajcn.2009.28947.

Chanda, M. L., and D. J. Levitin. "The neurochemistry of music." *Trends in Cognitive Sciences* 17 (2013): 179–193. doi:10.1016/j.tics.2013.02.007.

Chang, Larry. *Wisdom for the Soul: Five Millennia of Prescriptions for Spiritual Healing.* Washington, DC: Gnosophia Publishers, 2006.

"Chasing the Moon: Transcript, part two." *American Experience*, PBS, July 10, 2019. https://www.pbs.org/wgbh/americanexperience/films/chasing-moon/#transcript.

Chen, L. H., and C. Wu. "Gratitude enhances change in athletes' self-esteem: The moderating role of trust in coach." *Journal of Applied Sport Psychology* 26 (2014): 349–362. doi:10.1080/ 10413200.2014.889255.

Chen, X., J. C. Velez, C. Barbosa, et al. "Smoking and perceived stress in relation to short salivary telomere length among caregivers of children with disabilities." *Stress* 18, no. 1 (January 2015). doi:10.3109/10253890.2014.969704.

Cherkas, L. F., J. L. Hunkin, B. S. Kato, et al. "The association between physical activity in leisure time and leukocyte telomere length." *Archives of Internal Medicine* 168, no. 2 (January 2008): 154–158. doi:10.1001/archinternmed.2007.39.

Chetty, S., A. R. Friedman, K. Taravosh-Lahn, et al. "Stress and glucocorticoids promote oligodendrogenesis in the adult hippocampus." *Molecular Psychiatry* 19 (2014): 1275–1283. doi:10.1038/ mp.2013.190.

Choi, K., J. Kim, G. W. Kim, et al. "Oxidative stress-induced necrotic cell death via mitochondira-dependent [*sic*] burst of reactive oxygen species." *Current Neurovascular Research* 6, no. 4 (November 2009): 213–222.

Clark, A. E., P. Frijters, and M. A. Shields. "Relative income, happiness, and utility: An explanation for the Easterlin paradox and other puzzles." *Journal of Economic Literature* 46, no. 1 (March 2008): 95–144.

Clarke, T. C., P. M. Barnes, L. I. Black, et al. "Use of yoga, meditation, and chiropractors among U.S. adults aged 18 and over." *NCHS Data Brief* 325 (2018): 1–8.

Colzato, L. S., and A. Kibele. "How different types of meditation can enhance athletic performance depending on the specific sport skills." *Journal of Cognitive Enhancement* 1 (2017): 122–126.

Conklin, Q. A., A. D. Crosswell, C. D. Saron, et al. "Meditation, stress processes, and telomere biology." *Current Opinion in Psychology* 28 (August 2019): 92–101. doi:10.1016/j.copsyc.2018.11.009.

Connolly, S. L., T. B. Stoop, M. W. Logue, et al. "Posttraumatic stress disorder symptoms, temperament, and the pathway to cellular senescence." *Journal of Traumatic Stress* 31, no. 5 (October 2018): 676–686. doi:10.1002/jts.22325.

Corliss, J. "The many ways volunteering is good for your heart." *Harvard Health Publishing*, June 3, 2016.

Corporation for National and Community Service, Office of Research and Policy Development. *The Health Benefits of Volunteering: A Review of Recent Research.* Washington, DC, 2007.

Crick, F. *The Astonishing Hypothesis: The Scientific Search for the Soul.* London: Simon & Schuster, 1994.

Crous-Bou, M., T. T. Fung, J. Prescott, et al. "Mediterranean diet and telomere length in Nurses' Health Study: Population based cohort study." *British Medical Journal* 349 (2014). doi:10.1136/bmj.g6674.

Csíkszentmihályi, M. "Attention and the holistic approach to behavior." In *The Stream of Consciousness*, edited by K. S. Pope and J. L. Singer, 335–385. New York: Plenum, 1978.

Csíkszentmihályi, M. *Beyond Boredom and Anxiety.* San Francisco: Jossey-Bass, 2005.

Csíkszentmihályi, M. *Flow: The Psychology of Optimal Experience.* New York: Harper and Row, 1990.

Czapiński, J. "The economics of happiness and psychology of wealth." MPRA Paper 52897, University Library of Munich, Germany, 2013.

Dadvand, P., J. Pujol, D. Macià, et al. "The association between lifelong greenspace exposure and 3-dimensional brain magnetic resonance imaging in Barcelona schoolchildren." *Environmental Health Perspectives* (February 2018). doi:10.1289/EHP1876.

Darwin, Charles. *The Origin of Species.* New York: HarperCollins, 2011.

De Garbino, J. P. *Children's Health and the Environment: A Global Perspective.* Geneva: World Health Organization, 2004.

Della Casa, C., trans. *Upanisad.* Torino: UTET, 1976.

de Niet, G., B. Tiemens, B. Lendemeijer, et al. "Music-assisted relaxation to improve sleep quality: Meta-analysis." *Journal of Advanced Nursing* 65, no. 7 (July 2009): 1356–1364. doi:10.1111/j.1365-2648.2009.04982.x.

Dennett, C. "Key ingredients of the Mediterranean diet—the nutritious sum of delicious parts." *Today's Dietitian* 18, no. 5 (May 2006): 28.

Desbordes, G., L. T. Negi, T. W. Pace, et al. "Effects of mindful-attention and compassion meditation training on amygdala response to emotional stimuli in an ordinary, non-meditative state." *Frontiers in Human Neuroscience* 6 (November 2012). doi:10.3389/fnhum.2012.00292.

"Dhyana." *Encyclopedia Britannica*, last modified November 13, 2014. https://www.britannica.com/topic/dhyana.

Diener, E., R. E. Lucas, and C. N. Scollon. "Beyond the hedonic treadmill: Revising the adaptation theory of well-being." *American Psychologist* 61, no. 4 (May–June 2006): 305–314.

Diener, E., and S. Oishi. "Money and happiness: Income and subjective well-being across nations." In *Subjective Well-Being across Cultures*, edited by E. Diener and E. M. Suh, 185–218. Cambridge, MA: MIT Press, 2000.

Digdon, N., and A. Koble. "Effects of constructive worry, imagery distraction, and gratitude interventions on sleep quality: A pilot trial." *Applied Psychology: Health and Well-Being* 3 (2011): 193–206. doi:10.1111/j.1758–0854.2011.01049.x.

Dowlati, Y., N. Herrmann, W. Swardfager, et al. "A meta-analysis of cytokines in major depression." *Biological Psychiatry* 67, no. 5 (2010): 446–457. doi:10.1016/j.biopsych.2009.09.033.

Drury, S. S., K. Theall, M. M. Gleason, et al. "Telomere length and early severe social deprivation: Linking early adversity and cellular aging." *Molecular Psychiatry* 17, no. 7 (July 2012): 719–727. doi:10.1038/mp.2011.53.

Du, M., J. Prescott, P. Kraft, et al. "Physical activity, sedentary behavior, and leukocyte telomere length in women." *American Journal of Epidemiology* 175, no. 5 (2012): 414–422.

Earth Charter. "The Earth Charter." 2000. https://earthcharter.org/read-the-earth-charter/preamble/#:~:text=Universal%20Responsibility&text=We%20are%20at%20once%20citizens,and%20the%20larger%20living%20world.

Easterlin, R. A. "Does economic growth improve the human lot? Some empirical evidence." In *Nations and Households in Economic Growth*, edited by P. A. David and M. Reder, 89–125. New York: Academic Press, 1974. doi:10.1016/B978-0-12-205050-3.50008-7.

Easterlin, R. A. "Will raising the incomes of all increase the happiness of all?" *Journal of Economic Behavior and Organization* 27 (1995): 35–47.

Easwaran, E. *To Love Is to Know Me*. Tomales, CA: Nilgiri Press, 1993.

Edwards, S. "Love and the brain." Harvard Medical School, Spring 2015. https://hms.harvard.edu/news-events/publications-archive/brain/love-brain.

Ekman, P., and R. J. Davidson. "Voluntary smiling changes regional brain activity." *Psychological Science* 4, no. 5 (1993): 342–345.

Elger, C. E., A. D. Friederici, and C. Koch. "Das Manifest: Elf führende Neurowissenschaftler über Gegenwart und Zukunft der Hirnforschung." *Gehirn und Geist* 6 (2004): 30–37.

Emmons, R. A., and M. E. McCullough. "Counting blessings versus burdens: An experimental investigation of gratitude and subjective well-being in daily life." *Journal of Personality and Social Psychology* 84, no. 2 (2003): 377–389.

Emmons, R. A., and M. E. McCullough. *The Psychology of Gratitude*. New York: Oxford University Press, 2004.

Epel, E. S., E. H. Blackburn, J. Lin, et al. "Accelerated telomere shortening in response to life stress." *Proceedings of the National Academy of Sciences* 101, no. 49 (December 2004). doi:10.1073/pnas.0407162101.

Epel, E. S., and G. J. Lithgow. "Stress biology and aging mechanisms: Toward understanding the deep connection between adaptation to stress and longevity." *Journals of Gerontology. Series A, Biological Sciences and Medical Sciences* 69, suppl. 1 (June 2014): S10–S16. doi:10.1093/gerona/glu055.

Factor-Litvak, P., E. Susser, K. Kezios, et al. "Leukocyte telomere length in newborns: Implications for the role of telomeres in human disease." *Pediatrics* 137, no. 4 (2016). doi:10.1542/ peds.2015–3927.

Field, A. E., T. Byers, D. J. Hunter, et al. "Weight cycling, weight gain, and risk of hypertension in women." *American Journal of Epidemiology* 150, no. 6 (September 1999): 573–579.

Firebaugh, G., and L. M. Tach. "Relative income and happiness: Are Americans on a hedonic treadmill?" Paper presented at the hundredth annual meeting of the American Sociological Association, Philadelphia, August 13–16, 2005.

Fisher, H. *Why Him? Why Her?* New York: Henry Holt, 2009.

Fisher, H. E., H. D. Island, J. Rich, et al. "Four broad temperament dimensions: Description, convergent validation correlations, and comparison with the Big Five." *Frontiers in Psychology* 6 (2015):1098. doi:10.3389/fpsyg.2015.01098.

Fisher, H. E., J. Rich, H. D. Island, et al. "Four primary temperament dimensions in the process of mate choice." Poster presented at the annual meeting of the American Psychological Association, San Diego, August 14, 2010.

Fisher, H. E., J. Rich, H. D. Island, et al. "The second to fourth digit ratio: A measure of two hormonally-based temperament dimensions." *Personality and Individual Differences* 49, no. 7 (2010): 773–777. doi:10.1016/j.paid.2010.06.027.

Fitzpatrick, A. L., R. A. Kronmal, J. P. Gardner, et al. "Leukocyte telomere length and cardiovascular disease in the cardiovascular health study." *American Journal of Epidemiology* 165, no. 1 (January 2007): 14–21.

"5 of the best exercises you can ever do." *Harvard Health Publishing*, July 7, 2020. https:// www.health.harvard.edu/staying-healthy/5-of-the-best-exercises-you-can-ever-do.

Fondazione Umberto Veronesi. "Alzheimers." Accessed March 6, 2023. https://www .fondazioneveronesi.it/magazine/tools-della-salute/glossario-delle-malattie /alzheimer-2.

Fowler, J. H., and N. A. Christakis. "Dynamic spread of happiness in a large social network: Longitudinal analysis over 20 years in the Framingham Heart Study." *British Medical Journal* 337 (2008). doi:10.1136/bmj.a2338.

Frank, R. H. *Luxury Fever: Why Money Fails to Satisfy in an Era of Excess.* New York: Free Press, 1999.

Frederick, S. "Hedonic treadmill." In *Encyclopedia of Social Psychology*, edited by R. Vohs and K. Baumeister, 419–420. London: SAGE, 2007.

Fredrickson, B. L., M. A. Cohn, K. A. Coffey, et al. "Open hearts build lives: Positive emotions, induced through loving-kindness meditation, build consequential personal resources." *Journal of Personality and Social Psychology* 95, no. 5 (2008): 1045–1062. doi:10.1037/a0013262.

Fredrickson, B. L., K. M. Grewen, K. A. Coffey, et al. "A functional genomic perspective on human well-being." *Proceedings of the National Academy of Sciences* 110, no. 33 (2013). doi:10.1073/pnas.1305419110.

Freedman, J. "Choosing optimism: An interview with Martin EP Seligman." *EQ Life, Six Seconds*, November 10, 1999.

Fulton, H., R. Huisman, J. Murphet, et al. *Narrative and Media.* New York: Cambridge University Press, 2005. doi:10.1017/CBO9780511811760.

Gallai, N., J. M. Salles, J. Settele, et al. "Economic valuation of the vulnerability of world agriculture confronted with pollinator decline." *Economia Ecologica* 68, no. 3 (2009): 810–821.

Gaser, C., and G. Schlaug. "Brain structures differ between musicians and non-musicians." *Journal of Neuroscience* 23, no. 27 (October 2003): 9240–9245.

Gelli, Alessandro, Michele Cavallo, and Vito Ferri. "Dal respiro alla voce." Course material for Chair of Clinical Psycho-physiology Professor R. Venturini at the Sapienza University of Rome, n.d.

Gielen, M., G. J. Hageman, E. E. Antoniou, et al. "Body mass index is negatively associated with telomere length: A collaborative cross-sectional meta-analysis of 87 observational studies." *American Journal of Clinical Nutrition* 108, no. 3 (September 2018): 453–475. doi:10.1093/ajcn/nqy107.

Global Wellness Summit. *2019 Global Wellness Trends Report.* Miami, 2019.

Goethe, Johann Wolfgang von. *Wilhelm Meister's Apprenticeship.* Bk. 5. Berlin: Johann Friedrich Unger, 1795–1796.

Gordon, A. M., E. A. Impett, A. Kogan, et al. "To have and to hold: Gratitude promotes relationship maintenance in intimate bonds." *Journal of Personality and Social Psychology* 103, no. 2 (2012): 257–274.

Goyal, M., S. Singh, E. M. S. Sibinga, et al. "Meditation programs for psychological stress and well-being: A systematic review and meta-analysis." *JAMA Internal Medicine* 174, no. 3 (2014): 357–368. doi:10.1001/jamainternmed.2013.13018.

"Gratitude." *Merriam-Webster*, 2019. https://www.merriam-webster.com/dictionary/gratitude.

Griep, Y., L. Magnusson Hanson, T. Vantilborgh, et al. "Can volunteering in later life reduce the risk of dementia? A 5-year longitudinal study among volunteering and non-volunteering retired seniors." *PLOS ONE* (March 2017). doi:10.1371/journal.pone.0173885.

"Growing up in a Romanian orphanage." BBC, April 6, 2016.

Hagerty, M. R. "Social comparisons of income in one's community: Evidence from national surveys of income and happiness." *Journal of Personality and Social Psychology* 78, no. 4 (2000): 764–771. doi:10.1037/0022–3514.78.4.764.

Hagner, M. *Homo cerebralis: Der Wandel vom Seelenorgan zum Gehirn.* Frankfurt: Insel, 2000.

Harari, Y. N. *Da animali a dèi: Breve storia dell'umanità.* Translated by Giuseppe Bernardi. Milan: Bompiani, 2014.

Harper, Q., E. L. Worthington, B. G. Griffin, et al. "Efficacy of a workbook to promote forgiveness: A randomized controlled trial with university students." *Journal of Clinical Psychology* 70, no. 12 (2014): 1158–1169. doi:10.1002/jclp.22079.

Hartig, T. "Green space, psychological restoration, and health inequality." *Lancet* 372 (2008): 1614–1615. doi:10.1016/ S0140–6736(08)61669–4.

Hasenkamp, W., and L. W. Barsalou. "Effects of meditation experience on functional connectivity of distributed brain networks." *Frontiers in Human Neuroscience* 1 (March 2012). doi:10.3389/ fnhum.2012.00038.

Haybron, D. "Happiness." In *The Stanford Encyclopedia of Philosophy (Winter 2019 Edition),* edited by Edward N. Zalta. Stanford, CA: Stanford University, 2019. https:// plato.stanford.edu/archives/win2019/entries/happiness/.

Haycock, P. C., E. E. Heydon, S. Kaptoge, et al. "Leucocyte telomere length and risk of cardiovascular disease: Systematic review and meta-analysis." *British Medical Journal* 349, no. 8 (July 2014). doi:10.1136/ bmj.g4227.

"The health benefits of strong relationships." *Harvard Health Publishing,* August 6, 2019.

Heidegger, M., and H. S. Hisamatsu. "L'arte e il pensiero." In *L'Oriente di Heidegger,* edited by C. Saviani. Genova: Il Melangolo, 1998.

Hendrikx, P., M. P. Chauzat, M. Debin, et al. "Bee mortality and bee surveillance in Europe." *EFSA Supporting Publications* 6, no. 9 (December 2009).

Hernandez, R., M. L. Daviglus, L. Martinez, et al. "'¡Alegrate!'—a culturally adapted positive psychological intervention for Hispanics/Latinos with hypertension: Rationale, design, and methods." *Contemporary Clinical Trials Communications* 14 (2019). doi:10.1016/j.conctc.2019.100348.

Hilbrand, S., D. A. Coall, A. H. Meyer, et al. "A prospective study of associations among helping, health, and longevity." *Social Science and Medicine* 187 (August 2017): 109–117.

Hill, P. L., M. Allemand, and B. W. Roberts. "Examining the pathways between gratitude and self-rated physical health across adulthood." *Personality and Individual Differences* 54, no. 1 (2013): 92–96.

Hoge, E. A., M. M. Chen, E. Orr, et al. "Loving-kindness meditation practice associated with longer telomeres in women." *Brain, Behavior, and Immunity* 32 (August 2013): 159–163.

Holt-Lunstad, J., T. B. Smith, and J. B. Layton. "Social relationships and mortality risk: A meta-analytic review." *PLOS Medicine* 7, no. 7 (July 2020). doi:10.1371/journal .pmed.1000316.

Hölzel, B. K., J. Carmody, K. C. Evans, et al. "Stress reduction correlates with structural changes in the amygdala." *Social Cognitive and Affective Neuroscience* 5, no. 1 (March 2010): 11–17. doi:10.1093/scan/nsp034.

Hölzel, B. K., J. Carmody, M. Vangel, et al. "Mindfulness practice leads to increases in regional brain gray matter density." *Psychiatry Research* 191, no. 1 (January 2011): 36–43. doi:10.1016/j.pscychresns.2010.08.006.

Houben, J. M., H. J. Moonen, F. J. van Schooten, et al. "Telomere length assessment: Biomarker of chronic oxidative stress?" *Free Radical Biology and Medicine* 44, no. 3 (2008): 235–246. doi:10.1016/j.freeradbiomed.2007.10.001.

Houston, D. K., M. Cesari, L. Ferrucci, et al. "Association between vitamin D status and physical performance: The InCHIANTI study." *Journals of Gerontology: Series A, Biological Sciences and Medical Sciences* 62, no. 4 (April 2007): 440–446. doi:10.1093/ gerona/62.4.440.

Hu, F. B., R. J. Sigal, J. W. Rich-Edwards, et al. "Walking compared with vigorous physical activity and risk of type 2 diabetes in women: A prospective study." *JAMA* 282, no. 15 (August 1999): 1433–1439.

Huffman, J. C., C. A. Mastromauro, J. K. Boehm, et al. "Development of a positive psychology intervention for patients with acute cardiovascular disease." *Heart International* 6, no. 2 (2011). doi:10.4081/hi.2011.e14.

Hutcherson, C. A., E. M. Seppala, and J. J. Gross. "Loving-kindness meditation increases social connectedness." *Emotion* 8, no. 5 (2008): 720–724. doi:10.1037/ a0013237.

"Importance of sleep: Six reasons not to scrimp on sleep." *Harvard Health Publishing,* January 1, 2006.

INRAN. *Linee guida per una sana alimentazione italiana.* Ministry of Agricultural and Forestry Policies, last updated February 27, 2013. https://www.salute.gov.it /imgs/C_17_pubblicazioni_652_allegato.pdf.

IPBES. *The Assessment Report of the Intergovernmental Science-Policy Platform on Biodiversity and Ecosystem Services on Pollinators, Pollination and Food Production,* edited by S. G. Potts, V. L. Imperatriz-Fonseca, and H. T. Ngo. Bonn, Germany: Secretariat of the Intergovernmental Science-Policy Platform on Biodiversity and Ecosystem Services, 2016. https://doi.org/10.5281/zenodo.3402856.

IUCN. *Nearly One in 10 Wild Bee Species Face Extinction in Europe while the Status of More Than Half Remains Unknown—IUCN Report.* International Union for Conservation of Nature and Natural Resources, March 19, 2015. https://www.iucn.org /content/nearly-one-10-wild-bee-species-face-extinction-europe-while-status-more -half-remains-unknown-iucn-report.

Ivarsson, A., U. Johnson, M. B. Andersen, et al. "It pays to pay attention: A mindfulness-based program for injury prevention with soccer players." *Journal of Applied Sport Psychology* 27, no. 3 (2015): 319–334.

Jacobs, T. L., E. S. Epel, J. Lin, et al. "Intensive meditation training, immune cell telomerase activity, and psychological mediators." *Psychoneuroendocrinology* 36, no. 5 (June 2011): 664–681. doi:10.1016/j.psyneuen.2010.09.010.

James, P., J. E. Hart, R. F. Banay, et al. "Exposure to greenness and mortality in a nationwide prospective cohort study of women." *Environmental Health Perspectives* 124, no. 9 (September 2016). doi:10.1289/ehp.1510363.

James, William. "Pragmatism and common sense." Lecture 5 in *Pragmatism: A New Name for Some Old Ways of Thinking*, 63–75. New York: Longman Green, 1907.

Janssen, H. C., M. M. Samson, and H. J. Verhaar. "Vitamin D deficiency, muscle function, and falls in elderly people." *American Journal of Clinical Nutrition* 75, no. 4 (April 2002): 611–615. doi:10.1093/ajcn/75.4.611.

Jeanclos, E., N. J. Schork, K. O. Kyvik, et al. "Telomere length inversely correlates with pulse pressure and is highly familial." *Hypertension* 36, no. 2 (August 2000): 195–200.

Ji, L. L., M. C. Gomez-Cabrera, and J. Vina. "Exercise and hormesis: Activation of cellular antioxidant signaling pathway." *Annals of the New York Academy of Sciences* 1067 (May 2006): 425–435. doi:10.1196/annals.1354.061.

Jones, M. R., R. R. Fay, and A. N. Poppers, eds. *Music Perception.* New York: Springer, 2010.

Jorde, R., M. Sneve, Y. Figenschau, et al. "Effects of vitamin D supplementation on symptoms of depression in overweight and obese subjects: Randomized double-blind trial." *Journal of Internal Medicine* 264, no. 6 (October 2008): 599–609. doi:10.1111/j.13652796.2008.02008.x.

Judd, S., and V. Tangpricha. "Vitamin D deficiency and risk for cardiovascular disease." *Circulation* 117, no. 4 (January 2008): 503–511. doi:10.1161/CIRCULATIONAHA.107.706127.

Kabat-Zinn, J. *Full Catastrophe Living: Using the Wisdom of Your Body and Mind to Face Stress, Pain, and Illness.* New York: Delta, 2009.

Kabat-Zinn, J. "Mindfulness-based interventions in context: Past, present, and future." *Clinical Psychology: Science and Practice* 10 (2003): 144–156.

Kahneman, D., E. Diener, and N. Schwarz, eds. *Well-Being: The Foundations of Hedonic Psychology.* New York: Russell Sage Foundation, 1999.

Kanduri, C., P. Raijas, M. Ahvenainen, et al. "The effect of listening to music on human transcriptome." *PeerJ* 3 (2015). doi:10.7717/peerj.830.

Kashdan, T. B., G. Uswatte, and T. Julian. "Gratitude and hedonic and eudaimonic well-being in Vietnam War veterans." *Behavior Research and Therapy* 44, no. 2 (February 2006): 177–199. doi:10.1016/j.brat.2005.01.005.

Kaufman, K. A., C. R. Glass, and D. B. Arnkoff. "Evaluation of mindful sport performance enhancement (MSPE): A new approach to promote flow in athletes." *Journal of Clinical Sports Psychology* 4 (2009): 334–356.

Kaufman, M. "Meditation gives brain a charge, study finds." *Washington Post*, January 3, 2005.

Kawachi, I., G. A. Colditz, A. Ascherio, et al. "A prospective study of social networks in relation to total mortality and cardiovascular disease in men in the USA." *Journal of Epidemiology and Community Health* 50, no. 3 (June 1996): 245–251. doi:10.1136/jech.50.3.245.

Kennel, K. A., M. T. Drake, and D. L. Hurley. "Vitamin D deficiency in adults: When to test and how to treat." *Mayo Clinic Proceedings* 85, no. 8 (August 2010): 752–757. doi:10.4065/mcp.2010.0138.

Khalfa, S., S. D. Bella, M. Roy, et al. "Effects of relaxing music on salivary cortisol level after psychological stress." *Annals of the New York Academy of Sciences* 999 (November 2003): 374–376. doi:10.1196/annals.1284.045.

Khalsa, D. S. "Stress, meditation, and Alzheimer's disease prevention: Where the evidence stands." *Journal of Alzheimer's Disease* 48, no. 1 (2015): 1–12. doi:10.3233/JAD-142766.

Kim, E. S., S. W. Delaney, and L. D. Kubzansky. "Sense of purpose in life and cardiovascular disease: Underlying mechanisms and future directions." *Current Cardiology Reports* 21, no. 135 (October 2019). doi:10.1007/s11886-019-1222-9.

Kim, E. S., I. Kawachi, Y. Chen, et al. "Association between purpose in life and objective measures of physical function in older adults." *JAMA Network Open* 74, no. 10 (October 2017): 1039–1045. doi:10.1001/jamapsychiatry.2017.2145.

Kim, E. S., J. K. Sun, N. Park, et al. "Purpose in life and reduced incidence of stroke in older adults: 'The Health and Retirement Study.'" *Journal of Psychosomatic Research* 74, no. 5 (2013): 427–432. doi:10.1016/j.jpsychores.2013.01.013.

Kimura, M., Y. Gazitt, X. Cao, et al. "Synchrony of telomere length among hematopoietic cells." *Experimental Hematology* 38, no. 10 (October 2010): 854–859. doi:10.1016/j.exphem.2010.06.010.

Knight, J., and R. Gunatilaka. "Income, aspirations and the hedonic treadmill in a poor society." *Journal of Economic Behavior and Organization* 82 (2012). doi:10.1016/j.jebo.2011.12.005.

Koelsch, S., and T. Stegemann. "The brain and positive biological effects in healthy and clinical populations." In *Music, Health, and Wellbeing*, edited by R. A. R. MacDonald, G. Kreutz, and L. Mitchell, 436–456. Oxford: Oxford University Press, 2012.

Koenen, K. C., I. De Vivo, J. Rich-Edwards, et al. "Protocol for investigating genetic determinants of posttraumatic stress disorder in women from the Nurses' Health Study II." *BMC Psychiatry* 9, no. 29 (2009). doi:10.1186/1471-244X-9-29.

Konrath, S., A. Fuhrel-Forbis, A. Lou, et al. "Motives for volunteering are associated with mortality risk in older adults." *Health Psychology* 31, no. 1 (2012): 87–96. doi:10.1037/a0025226.

Korb, A. "The grateful brain." *Psychology Today*, November 20, 2012. https://www
.psychologytoday.com/au/blog/prefrontal-nudity/201211/the-grateful-brain.

Kroenke, C. H., L. D. Kubzansky, E. S. Schernhammer, et al. "Social networks, social
support, and survival after breast cancer diagnosis." *Journal of Clinical Oncology* 24,
no. 7 (March 2006): 1105–1111. doi:10.1200/JCO.2005.04.2846.

Kubzansky, L. D., D. Sparrow, P. Vokonas, et al. "Is the glass half empty or
half full? A prospective study of optimism and coronary heart disease in the
normative aging study." *Psychosomatic Medicine* 63, no. 6 (2001): 910–916.
doi:10.1097/00006842-200111000-00009.

Kyeong, S., J. Kim, D. J. Kim, et al. "Effects of gratitude meditation on neural network
functional connectivity and brain-heart coupling." *Scientific Reports* 7. doi:10.1038/
s41598-017-05520-9.

La 'Carta dei Valori della Dieta Mediterranea UNESCO' patrimonio culturale imma-
teriale dell'umanità." Ministero delle politiche agricole alimentari e forestali and
Unitelma Sapienza, 2014. https://www.obesityday.org/usr_files/biblioteca/Carta
_valori_dieta_mediterranea.pdf.

Laden, F., and J. Johnston. "Green spaces and health." National Institute of Environmental Health Sciences, webinar, September 26, 2016. https://www.niehs.nih.gov
/research/supported/translational/peph/webinars/green_spaces/index.cfm.

Landau, E. "Singing therapy helps stroke patients regain language." CNN, February
22, 2010.

Lau, R. W., and S. T. Cheng. "Gratitude lessens death anxiety." *European Journal of
Ageing* 8, no. 3 (2011): 169.

Lautenbach, S., R. Seppelt, J. Liebscher, et al. "Spatial and temporal trends of global
pollination benefit." *PLOS ONE* (April 2012). doi:10.1371/journal.pone.0035954.

Lawrence, E. M., R. G. Rogers, and T. Wadsworth. "Happiness and longevity in
the United States." *Social Science and Medicine* 145 (November 2015): 115–119.
doi:10.1016/j.socscimed.2015.09.020.

Lazar, S. W., G. Bush, R. L. Gollub, et al. "Functional brain mapping of the relaxation response and meditation." *Neuroreport* 11, no. 7 (May 2000): 1581–1585.

Lee, J. H. "The effects of music on pain: A meta-analysis." *Journal of Music Therapy*
53, no. 4 (2016): 430–477. doi:10.1093/jmt/thw012.

Lee, L. O., P. James, E. S. Zevon, et al. "Optimism is associated with exceptional
longevity in 2 epidemiologic cohorts of men and women." *Proceedings of the National
Academy of Sciences* 116, no. 37 (2019): 18357–18362. doi:10.1073/pnas.1900712116.

Le Nguyen, K. D., J. Lin, S. B. Algoe, et al. "Loving-kindness meditation slows biological aging in novices: Evidence from a 12-week randomized controlled trial."
Psychoneuroendocrinology 108 (2019): 20–27.

Liang, G., E. Schernhammer, L. Qi, et al. "Associations between rotating night
shifts, sleep duration, and telomere length in women." *PLOS ONE* 6, no. 8 (2011).
doi:10.1371/journal.pone.0023462.

Lin, W.-Y., C.-C. Chan, Y.-L. Liu, et al. "Performing different kinds of physical exercise differentially attenuates the genetic effects on obesity measures: Evidence from 18,424 Taiwan Biobank participants." *PLOS Genetics* 15, no. 8 (August 2019). doi:10.1371/journal.pgen.1008277.

Linnemann, A., B. Ditzen, J. Strahler, et al. "Music listening as a means of stress reduction in daily life." *Psychoneuroendocrinology* 60 (October 2015): 82–90. doi:10.1016/j.psyneuen.2015.06.008.

Liu, S., M. J. Stampfer, F. B. Hut, et al. "Whole-grain consumption and risk of coronary heart disease: Results from the Nurses' Health Study." *American Journal of Clinical Nutrition* 70, no. 3 (September 1999): 412–419.

Loomba, R. S., P. H. Shah, S. Chandrasekar, et al. "Effects of music on systolic blood pressure, diastolic blood pressure, and heart rate: A meta-analysis." *Indian Heart Journal* 64, no. 3 (2012): 309–313. doi:10.1016/S0019–4832(12)60094–7.

Lutz, A., H. A. Slagter, J. D. Dunne, et al. "Attention regulation and monitoring in meditation." *Trends in Cognitive Sciences* 12, no. 4 (2008): 163–169.

Lykken, D. T. "'Beyond the hedonic treadmill: Revising the adaptation theory of well-being': Comment on Diener, Lucas, and Scollon (2006)." *American Psychologist* 62, no. 6 (2007): 611–612. doi:10.1037/0003–066X62.6.611.

Lyubomirsky, S., and M. Della Porta. "Boosting happiness, buttressing resilience: Results from cognitive and behavioral interventions." In *Handbook of Adult Resilience: Concepts, Methods, and Applications*, edited by J. W. Reich, A. J. Zautra, and J. Hall. New York: Guilford Press, 2012.

Mahony, J., and S. J. Hanrahan. "A brief educational intervention using acceptance and commitment therapy: Four injured athletes experiences." *Journal of Clinical Sport Psychology* 5 (2011): 252–273.

Mandela, Nelson. Preface to *Mandela's Way: Lessons on Life, Love, and Courage*, by Richard Stengel. New York: Crown Publishers, 2009.

Manson, J. E., F. B. Hu, J. W. Rich-Edwards, et al. "A prospective study of walking as compared with vigorous exercise in the prevention of coronary heart disease in women." *New England Journal of Medicine* 341, no. 9 (August 1999): 650–658.

Marchi, I. *Fiori, mine e alcune domande.* Turin: Sillabe di Sale, 2015.

Marrone, C. "Che cosa succede al tuo corpo quando dormi." *Corriere della Sera,* September 14, 2018.

Marsh, L. "Hayek: Cognitive scientist Avant la Lettre." *Advances in Austrian Economics* 13 (2010): 115–155.

Mathieu, C., C. Gysemans, A. Giulietti, et al. "Vitamin D and diabetes." *Diabetologia* 48, no. 7 (2005): 1247–1257. doi:10.1007/s00125-005-1802-7.

Matthieu, M. M., K. A. Lawrence, and E. Robertson-Blackmore. "The impact of a civic service program on biopsychosocial outcomes of post 9/11 U.S. military veterans." *Psychiatry Research* 248 (February 2017): 111–116. doi:10.1016/j.psych res.2016.12.028.

McGrath, M., J. Y. Wong, D. Michaud, et al. "Telomere length, cigarette smoking, and bladder cancer risk in men and women." *Cancer Epidemiology, Biomarkers, and Prevention* 16, no. 4 (April 2007): 815–819.

McMahon, D. M. "The pursuit of happiness in history." In *The Science of Subjective Well-Being*, edited by M. Eid and R. J. Larsen, 80–93. New York: Guilford Press, 2008.

McPhee, J. S., D. P. French, D. Jackson, et al. "Physical activity in older age: Perspectives for healthy ageing and frailty." *Biogerontology* 17 (2016): 567–580. doi:10.1007/s10522-016-9641-0.

Mejia, M. "Harvard's longest study of adult life reveals how you can be happier and more successful." CNBC, March 20, 2018.

Menon, V., and D. J. Levitin. "The rewards of music listening: Response and physiological connectivity of the mesolimbic system." *Neuroimage* 28 (2005): 175–184. doi:10.1016/j.neuroimage.2005.05.053.

Menotti, A., and P. E. Puddu. "How the Seven Countries Study contributed to the definition and development of the Mediterranean diet concept: A 50-year journey." *Nutrition, Metabolism, and Cardiovascular Disease* 25, no. 3 (March 2015): 245–252. doi:10.1016/j.numecd.2014.12.001.

Mineo, L. "With mindfulness, life's in the moment: Those who learn its techniques often say they feel less stress, think clearer." *Harvard Gazette*, April 1, 2018.

Ministero della Salute. "Obesità." Italian Ministry of Health, March 2, 2021. https://www.salute.gov.it/portale/nutrizione/dettaglioContenutiNutrizione.jsp?lingua=italiano&id=5510&area=nutrizione&menu=croniche.

Mischel, W., E. B. Ebbesen, and A. Raskoff Zeiss. "Cognitive and attentional mechanisms in delay of gratification." *Journal of Personality and Social Psychology* 21, no. 2 (1972): 204–218. https://doi.org/10.1037/h0032198.

Morvillo, C. "La scienza dell'amore, così il cervello reagisce ai sentimenti." *Corriere della Sera*, February 7, 2020. https://www.corriere.it/moda/san-valentino/notizie/scienza-dell-amore-cosi-cervello-reagisce-sentimenti-2ca887e0-49e2-11ea-8e62fcd8bfe20a1c.shtml.

Mozes Kor, E. *Ad Auschwitz ho imparato il perdono.* Milan: Sperling and Kupfer, 2017.

Muehsam, D., and C. Ventura. "Life rhythm as a symphony of oscillatory patterns: Electromagnetic energy and sound vibration modulates gene expression for biological signaling and healing." *Global Advances in Health and Medicine* 3, no. 2 (2014): 40–55. doi:10.7453/gahmj. 2014.008.

"Music and health." *Harvard Health Publishing*, July 2011.

Nakamura, J., and M. Csíkszentmihályi. "The concept of flow." In *Handbook of Positive Psychology*, edited by C. R. Snyder and S. J. Lopez, 89–105. New York: Oxford University Press, 2005.

National Institute of Neurological Disorders and Stroke. *Brain Basics: Understanding Sleep.* Rockville, MD: National Institute of Neurological Disorders and Stroke, August 2017. https://catalog.ninds.nih.gov/publications/understanding-sleep-brain-basics.

Nawrot, T. S., J. A. Staessen, J. P. Gardner, et al. "Telomere length and possible link to X chromosome." *Lancet* 363, no. 9408 (February 2004): 507–510.

Nelson, C. A., C. H. Zeanah, N. A. Fox, et al. "Cognitive recovery in socially deprived young children: The Bucharest Early Intervention Project." *Science* 318, no. 5858 (2007): 1937–1940. doi:10.1126/science.1143921.

Nieto, A., S. P. M. Roberts, J. Kemp, et al. *European Red List of Bees*. Luxembourg: Publication Office of the European Union, 2014. doi:10.2779/77003.

Nussbaum, M. C., and A. Sen, eds. *The Quality of Life*. Oxford: Clarendon Press, 1993.

Oikawa, S., and S. Kawanishi. "Site-specific DNA damage at GGG sequence by oxidative stress may accelerate telomere shortening." *FEBS Letters* 453, no. 3 (June 1999): 365–368.

Okereke, O., J. Prescott, J. Wong, et al. "High phobic anxiety is related to lower leukocyte telomere length in women." *PLOS ONE* 7 (2012). doi:10.1371/journal .pone.0040516.

Okuda, K., A. Bardeguez, J. P. Gardner, et al. "Telomere length in the newborn." *Pediatric Research* 52, no. 3 (September 2002): 377–381.

Oman, D., C. E. Thoresen, and K. McMahon. "Volunteerism and mortality among the community-dwelling elderly." *Journal of Health Psychology* 4, no. 3 (May 1999): 301–316. doi:10.1177/135910539900400301.

"1 in 3 adults don't get enough sleep." CDC Newsroom Releases, February 18, 2016.

Opresko, P. L., J. Fan, S. Danzy, et al. "Oxidative damage in telomeric DNA disrupts recognition by TRF1 and TRF2." *Nucleic Acids Research* 33, no. 4 (February 2005): 1230–1239.

"Optimism and your health." *Harvard Health Publishing*, May 1, 2008.

Pace, T. W., L. T. Negi, D. D. Adame, et al. "Effect of compassion meditation on neuroendocrine, innate immune and behavioral responses to psychosocial stress." *Psychoneuroendocrinology* 34, no. 1 (January 2009): 87–98. doi:10.1016/j.psyneuen .2008.08.011.

Parker, G. B., H. Brotchie, and R. K. Graham. "Vitamin D and depression." *Journal of Affective Disorders* 208 (2017): 56–61.

Pascal, Blaise. *Pensées*. Paris: Guillaume Desprez, 1670.

Pegg Frates, E. "Time spent in 'green' places linked with longer life in women." *Harvard Health Publishing*, March 9, 2017.

Peng, C. K., J. E. Mietus, Y. Liu, et al. "Exaggerated heart rate oscillations during two meditation techniques." *International Journal of Cardiology* 70, no. 2 (July 1990): 101–107.

Perez-de-Albeniz, A., and J. Holmes. "Meditation: Concepts, effects and uses in therapy." *International Journal of Psychotherapy* 5, no. 1 (March 2000): 49–59. doi:10.1080/13569080050020263.

Phillips, N. K., C. L. Hammen, P. A. Brennan, et al. "Early adversity and the prospective prediction of depressive and anxiety disorders in adolescents." *Journal of Abnormal Child Psychology* 33, no. 1 (2005): 13–24. doi:10.1007/s10802-005-0930-3.

Powell, A. "Can happiness lead toward health? At conference, new Harvard center explores ties between those major life factors." *Harvard Gazette*, December 5, 2016.

Powell, A. "When science meets mindfulness: Researchers study how it seems to change the brain in depressed patients." *Harvard Gazette*, April 9, 2018.

Prescott, J., M. Du, J. Y. Y. Wong, et al. "Paternal age at birth is associated with offspring leukocyte telomere length in the Nurses' Health Study." *Human Reproduction* 27, no. 12 (December 2012): 3622–3631. doi:10.1093/humrep/des314.

Preziosi, E. *Corso di Meditazione di Mindfulness: Conosco, conduco, calmo il mio pensare (con 8 brani per la pratica da scaricare online).* 2nd ed. Milan: FrancoAngeli, 2016.

Proto, E., and A. Rustichini. "A reassessment of the relationship between GDP and life satisfaction." *PLOS ONE* 8, no. 11 (2013). doi:10.1371/journal.pone.0079358.

Purcell, N., B. J. Griffin, K. Burkman, et al. "'Opening a door to a new life': The role of forgiveness in healing from moral injury." *Frontiers in Psychiatry* 9 (October 2018). doi:10.3389/fpsyt.2018.00498.

Puterman, E., J. Lin, E. Blackburn, et al. "The power of exercise: Buffering the effect of chronic stress on telomere length." *PLOS ONE* 5, no. 5 (2012). doi:10.1371/journal.pone.0010837.

Radak, Z., H. Y. Chung, and S. Goto. "Systemic adaptation to oxidative challenge induced by regular exercise." *Free Radical Biology and Medicine* 44, no. 2 (2008): 153–159. doi:10.1016/j.freeradbiomed.2007.01.029.

Rauscher, F. H., G. L. Shaw, and C. N. Ky. "Music and spatial task performance." *Nature* 365, no. 6447 (1993): 611. doi:10.1038/365611a0.

Renzaho, A. M. N., B. Houng, J. Oldroyd, et al. "Stressful life events and the onset of chronic diseases among Australian adults: Findings from a longitudinal survey." *European Journal of Public Health* 24, no. 1 (2014): 57–62. doi:10.1093/eurpub/ckt007.

Ribeiro, A. I., C. Tavares, A. Guttentag, et al. "Association between neighbourhood green space and biological markers in school-aged children: Findings from the Generation XXI birth cohort." *Environment International* 132 (November 2019). doi:10.1016/j.envint.2019.105070.

Roberts, A. L., K. C. Koenen, Q. Chen, et al. "Posttraumatic stress disorder and accelerated aging: PTSD and leukocyte telomere length in a sample of civilian women." *Depression and Anxiety* 34, no. 5 (2017): 391–400. doi:10.1002/da.22620.

Rockhill, B., W. C. Willett, D. J. Hunter, et al. "A prospective study of recreational physical activity and breast cancer risk." *Archives of Internal Medicine* 159, no. 19 (October 1999): 2290–2296. doi:10.1001/archinte.159.19.2290.

Rode, L., B. G. Nordestgaard, and S. E. Bojesen. "Peripheral blood leukocyte telomere length and mortality among 64,637 individuals from the general population."

Journal of the National Cancer Institute 107, no. 6 (April 2015): djv074. doi:10.1093/jnci/djv074.

Rosenkranz, M. A., D. C. Jackson, K. M. Dalton, et al. "Affective style and in vivo immune response: Neurobehavioral mechanisms." *Proceedings of the National Academy of Sciences* 100 (2003): 11148–11152.

Rowland, L., and O. S. Curry. "A range of kindness activities boost happiness." *Journal of Social Psychology* 159, no. 3 (2019): 340–343. doi:10.1080/00224545.2018.1469461.

Rozanski, A., C. Bavishi, L. D. Kubzansky, et al. "Association of optimism with cardiovascular events and all-cause mortality: A systematic review and meta-analysis." *JAMA Network Open* 2, no. 9 (2019). doi:10.1001/jamanetworkopen.2019.12200.

Ryan, R. M., and E. L. Deci. "On happiness and human potentials: A review of research on hedonic and eudaimonic well-being." *Annual Review of Psychology* 52 (2001): 141–166. doi:10.1146/annurev.psych.52.1.141.

Ryff, C. D. "Psychological well-being in adult life." *Current Directions in Psychological Science* 4, no. 4 (1995): 99–104. doi:10.1111/1467-8721.ep10772395.

Ryff, C. D., and B. Singer. "The contours of positive human health." *Psychological Inquiry* 9 (1998): 1–28.

Ryff, C. D., and B. Singer. "Interpersonal flourishing: A positive health agenda for the new millennium." *Personality and Social Psychology Review* 4 (2000): 30–44.

Sacks, F. M. "Dietary fat, the Mediterranean diet, and health." *American Journal of Medicine* 113, suppl. 9B (December 2002): 1S–4S. doi:10.1016/ s0002–9343(01)00985–8.

Scheier, M. F., C. S. Carver, and M. W. Bridges. "Distinguishing optimism from neuroticism (and trait anxiety, self-mastery, and self-esteem): A re-evaluation of the Life Orientation Test." *Journal of Personality and Social Psychology* 67, no. 6 (1994): 1063–1078.

Schmeller, D. S., J. Niemelä, and P. Bridgewater. "The Intergovernmental Science-Policy Platform on Biodiversity and Ecosystem Services (IPBES): Getting involved." *Biodiversity and Conservation* 26 (2017): 2271–2275.

Schulte, B. "Harvard neuroscientist: Meditation not only reduces stress, here's how it changes your brain." *Washington Post*, May 26, 2015.

Schweickart, Russell. "No frames, no boundaries: Connecting with the whole planet—from space." *Rediscovering the North American Vision: Roots and Renewal (IC#3)* (Summer 1983). https://www.context.org/iclib/ic03/schweick/.

Scoglio, A. A. J., D. A. Rudat, D. Garvert, et al. "Self-compassion and responses to trauma: The role of emotion regulation." *Journal of Interpersonal Violence* (December 2015): 1–21. doi:10.1177/0886260515622296.

Segerstrom, S. C., and G. E. Miller. "Psychological stress and the human immune system: A meta-analytic study of 30 years of inquiry." *Psychological Bulletin* 130, no. 4 (July 2004): 601–630.

Seligman, M. "Learned optimism test." 2006. https://core-docs.s3.amazonaws.com/documents/asset/uploaded_file/1251357/OptimismQuiz-1.pdf.

Seligman, Martin E. P. *Learned Optimism*. New York: A. A. Knopf, 1991.

Selye, Hans. *The Stress of Life*. Rev. ed. New York: McGraw-Hill, 1976.

Shaji, J., S. K. Verma, and G. L. Khanna. "The effect of mindfulness meditation on HPA-axis in pre-competition stress in sports performance of elite shooters." *National Journal of Integrated Research in Medicine* 2, no. 3 (2011): 15–21.

Shakespeare, W. *La dodicesima notte*. Translated and edited by Agostino Lombardo. Milan: Feltrinelli, 1993.

Shapero, B. G., J. Greenberg, P. Pedrelli, et al. "Mindfulness-based interventions in psychiatry." *Focus (American Psychiatric Publishing)* 16, no. 1 (2018): 32–39. doi:10.1176/appi.focus.20170039.

Sharot, T., C. W. Korn, and R. J. Dolan. "How unrealistic optimism is maintained in the face of reality." *Nature Neuroscience* 14, no. 11 (October 2011): 1475–1479. doi:10.1038/nn.2949.

Shen, Owen, et al. "Death: Reality vs reported." 2018. Reported in Hannah Ritchie, Fiona Spponer, and Max Roser. "Causes of death." Our World in Data, February 2018, last updated December 2019. https://ourworldindata.org/causes-of-death.

Sherwood, C., D. Kneale, and B. Bloomfield. "The way we are now: The state of the UK's relationships." *Relate*, August 2014.

Shetty, Jay. *Think Like a Monk*. New York: Simon & Schuster, 2020.

Shwartz, M. "We've evolved to be smart enough to make ourselves sick: Robert Sapolsky discusses physiological effects of stress." *Stanford News*, March 7, 2007.

Silk, J. B., J. C. Beehner, T. J. Bergman, et al. "Strong and consistent social bonds enhance the longevity of female baboons." *Current Biology* 20, no. 15 (August 2010): 1359–1361. doi:10.1016/j.cub.2010.05.067.

Simon, N. M., J. W. Smoller, K. L. McNamara, et al. "Telomere shortening and mood disorders: Preliminary support for a chronic stress model of accelerated aging." *Biological Psychiatry* 60, no. 5 (2006): 432–435. doi:10.1016/j.biopsych.2006.02.004.

Sneed, R. S., and S. Cohen. "A prospective study of volunteerism and hypertension risk in older adults." *Psychology and Aging* 28, no. 2 (2013): 578–586. doi:10.1037/a0032718.

Sofi, F., R. Abbate, G. F. Gensini, et al. "Accruing evidence on benefits of adherence to the Mediterranean diet on health: An updated systematic review and meta-analysis." *American Journal of Clinical Nutrition* 92, no. 5 (November 2010): 1189–1196.

Sofi, F., F. Cesari, R. Abbate, et al. "Adherence to Mediterranean diet and health status: Meta-analysis." *British Medical Journal* 337 (September 2008). doi:10.1136/bmj.a1344.

Solberg, E. E., F. Ingjer, A. Holen, et al. "Stress reactivity to and recovery from a standardized exercise bout: A study of 31 runners practicing relaxation techniques." *British Journal of Sports Medicine* 34 (2000): 268–272.

Steenstrup, T., J. V. Hjelmborg, L. H. Mortensen, et al. "Leukocyte telomere dynamics in the elderly." *European Journal of Epidemiology* 28, no. 2 (February 2013): 181–187. doi:10.1007/s10654-013-9780-4.

Steptoe, A. "Happiness and health." *Annual Review of Public Health* 40 (2019): 339–359. doi:10.1146/annurev-publhealth-040218-044150.

Sullivan, Walter. "The Einstein Papers: A man of many parts." *New York Times*, March 29, 1972.

Sun, Q., L. Shi, J. Prescott, et al. "Healthy lifestyle and leukocyte telomere length in U.S. women." *PLOS ONE* 7, no. 5 (May 2012).

Suzuki, David. "Declaration of interdependence." David Suzuki Foundation. Presented at the United Nations' Earth Summit, Rio de Janeiro, 1992. https://suzukiel ders.org/declaration-of-interdependence/.

Swann, C., R. J. Keegan, D. Piggott, et al. "A systematic review of the experience, occurrence, and controllability of flow states in elite sport." *Psychology of Sport and Exercise* 13 (2012): 807–819.

Tang, Y. Y., Y. Ma, Y. Wang, et al. "Short-term meditation training improves attention and self-regulation." *Proceedings of the National Academy of Sciences* 104, no. 43 (August 2007). doi:10.1073/pnas.0707678104.

Teng, X. F., M. Y. M. Wong, and Y. T. Zhang. "The effect of music on hypertensive patients." *Annual International Conference of the IEEE EMBS* (2007): 4649–4651. doi:10.1109/IEMBS.2007.4353376.

Tetzner, J., and M. Becker. "Think positive? Examining the impact of optimism on academic achievement in early adolescents." *Journal of Personality* (March 2017). doi:org/10.1111/jopy. 12312.

"Tips to help you reach your exercise and weight loss goals." *Harvard Health Publishing*, August 7, 2012.

Trappe, H.-J., and G. Voit. "The cardiovascular effect of musical genres." *Deutsches Arzteblatt International* 113, no. 20 (May 2016): 347–352. doi:10.3238/arztebl.2016.0347.

Trew, J. L., and L. E. Alden. "Kindness reduces avoidance goals in socially anxious individuals." *Motivation and Emotion* (2015). doi:10.1007/s11031015-9499-5.

Trichopoulou, A. "Diversity v. globalization: Traditional foods at the epicentre." *Public Health Nutrition* 15, no. 6 (February 2012): 951–954. doi:10.1017/S1368980012000304.

Trichopoulou, A., T. Costacou, C. Bamia, et al. "Adherence to a Mediterranean diet and survival in a Greek population." *New England Journal of Medicine* 348 (2003): 2599–2608. doi:10.1056/NEJMoa025039.

Trichopoulou, A., E. Vasilopoulou, P. Hollman, et al. "Nutritional composition and flavonoid content of edible wild greens and green pies: A potential rich source of antioxidant nutrients in the Mediterranean diet." *Food Chemistry* 70 (2000): 319–323.

Trombetti, A., M. Hars, F. R. Herrmann, et al. "Effect of music-based multitask training on gait, balance, and fall risk in elderly people: A randomized controlled trial." *Archives of Internal Medicine* 171, no. 6 (March 2011): 525–533. doi:10.1001/archinternmed.2010.446.

US Department of Health and Human Services. *2008 Physical Activity Guidelines for Americans: Be Active, Healthy, and Happy!* Washington, DC, 2008.

Vance, M. C., E. Bui, S. S. Hoeppner, et al. "Prospective association between major depressive disorder and leukocyte telomere length over two years." *Psychoneuroendocrinology* 90 (April 2018): 157–164. doi:10.1016/j.psyneuen.2018.02.015.

van der Sluijs, J. P. "Insect decline, an emerging global environmental risk." *Current Opinion in Environmental Sustainability* 46 (October 2020): 39–42. https://doi.org/10.1016/j.cosust.2020.08.012.

VanderWeele, T. J. "Is forgiveness a public health issue?" *American Journal of Public Health* 108, no. 2 (February 2018). doi:10.2105/AJPH.2017.304210.

Vasan, R. S., S. Demissie, M. Kimura, et al. "Association of leukocyte telomere length with circulating biomarkers of the renin-angiotensin-aldosterone system: The Framingham Heart Study." *Circulation* 117, no. 9 (March 2008): 1138–1144. doi:10.1161/CIRCULATIONAHA.107.731794.

Vasilopoulou, E., and A. Trichopoulou. "Green pies: The flavonoid rich Greek snack." *Food Chemistry* 126 (2011): 855–858.

Venkatesh, S., T. R. Raju, Y. Shivani, et al. "A study of structure of phenomenology of consciousness in meditative and non-meditative states." *Indian Journal of Physiology and Pharmacology* 41, no. 2 (April 1997): 149–153.

von Zglinicki, T. "Oxidative stress shortens telomeres." *Trends in Biochemical Sciences* 27, no. 7 (July 2002): 339–344.

Wada, K., J. T. Howard, P. McConnell, et al. "A molecular neuroethological approach for identifying and characterizing a cascade of behaviorally regulated genes." *Proceedings of the National Academy of Sciences* 103 (2006): 15212–15217. doi:10.1073/pnas.0607098103.

Wade, N. G., W. T. Hoyt, J. E. Kidwell, et al. "Efficacy of psychotherapeutic interventions to promote forgiveness: A meta-analysis." *Journal of Consulting and Clinical Psychology* 82, no. 1 (February 2014): 154–170. doi:10.1037/a0035268.

Waldinger, R. J., S. Cohen, M. S. Schulz, et al. "Security of attachment to spouses in late life: Concurrent and prospective links with cognitive and emotional well-being." *Clinical Psychological Science* 3, no. 4 (June 2015): 516–529. doi:10.1177/2167702614541261.

Waldinger, R. J., and M. S. Schulz. "What's love got to do with it? Social functioning, perceived health, and daily happiness in married octogenarians." *Psychology and Aging* 25, no. 2 (June 2010): 422–431. doi:10.1037/a0019087.

Wallace, R. K. "Physiological effects of transcendental meditation." *Science* 167 (1970): 1751–1754. doi:10.1126/science.167.3926.1751.

Wallace, R. K., H. Benson, and A. F. Wilson. "A wakeful hypometabolic physiologic state." *American Journal of Physiology* 221 (September 1971): 795–799. doi:10.1152/ajplegacy.1971.221.3.795.

Waterman, A. S. "On the importance of distinguishing hedonia and eudaimonia when contemplating the hedonic treadmill." *American Psychologist* 62, no. 6 (2007): 612–613. doi:10.1037/0003066X62.6.612.

Watson, D. *The Wordsworth Dictionary of Musical Quotations.* Edinburgh: Wordsworth Reference, 1991.

Watson, S. "Volunteering may be good for body and mind." *Harvard Health Publishing,* June 26, 2013.

White, F. *The Overview Effect: Space Exploration and Human Evolution.* New York: Houghton Mifflin, 1987.

White, J. M. "Effects of relaxing music on cardiac autonomic balance and anxiety after acute myocardial infarction." *American Journal of Critical Care* 8, no. 4 (July 1999): 220–230.

White, M. P., I. Alcock, J. Grellier, et al. "Spending at least 120 minutes a week in nature is associated with good health and wellbeing." *Nature* 9 (2019). doi:10.1038/s41598-019-44097-3.

Williams, L., and M. Bartlett. "Warm thanks: Gratitude expression facilitates social affiliation in new relationships via perceived warmth." *Emotion* 15. doi:10.1037/emo0000017.

Wolkowitz, O. M., E. S. Epel, V. I. Reus, et al. "Depression gets old fast: Do stress and depression accelerate cell aging?" *Depression and Anxiety* 27, no. 4 (2010): 327–338. doi:10.1002/da.20686.

Wolkowitz, O. M., S. H. Mellon, E. S. Epel, et al. "Leukocyte telomere length in major depression: Correlations with chronicity, inflammation and oxidative stress—preliminary findings." *PLOS ONE* 6, no. 3 (March 2011). doi:10.1371/journal.pone.0017837.

Wong, J. Y. Y., I. De Vivo, X. Lin, et al. "Cumulative $PM_{2.5}$ exposure and telomere length in workers exposed to welding fumes." *Journal of Toxicology and Environmental Health, Part A* 77, no. 8 (2014): 441–455. doi:10.1080/15287394.2013.875497.

Wong, Y. J., J. Owen, N. T. Gabana, et al. "Does gratitude writing improve the mental health of psychotherapy clients? Evidence from a randomized controlled trial." *Psychotherapy Research* 28, no. 2 (2018): 192–202.

Woo, J., N. L. Tang, E. Suen, et al. "Telomeres and frailty." *Mechanism of Ageing and Development* 129, no. 11 (November 2008): 642–648. doi:10.1016/j.mad.2008.08.003.

World Health Organization, Regional Office for Europe. "WHO technical meeting on sleep and health." Bonn, Germany, January 22–24, 2004. https://apps.who.int/iris/handle/10665/349782.

Yang, B.-Y., X.-W. Zeng, I. Markevych, et al. "Association between greenness surrounding schools and kindergartens and attention-deficit/hyperactivity disorder in children in China." *JAMA Network Open* 18 (December 2019). doi:10.1001/jama networkopen.2019.17862.

Yang, J., A. Bakshi, Z. Zhu, et al. "Genetic variance estimation with imputed variants finds negligible missing heritability for human height and body mass index." *Nature Genetics* 47, no. 10 (October 2015). doi:10.1038/ng.3390.

Yeung, J. W. K., Z. Zhang, and T. Y. Kim. "Volunteering and health benefits in general adults: Cumulative effects and forms." *BMC Public Health* 18 (July 2017). doi:10.1186/s12889-017-4561-8.

Yogananda, Paramahansa. *Man's Eternal Quest, and Other Talks*. New Delhi: Oxford & IBH, 1975.

Zappalà, Alberto. *Riza Scienze*, December 1998.

Zee Ma, Y., and Y. Zhang. "Resolution of the happiness–income paradox." *Social Indicators Research* 119, no. 2 (2014): 705–721. www.jstor.org/stable/24721450.

Zollars, I., T. I. Poirier, and J. Pailden. "Effects of mindfulness meditation on mindfulness, mental well-being, and perceived stress." *Currents in Pharmacy Teaching and Learning* 11 (2019): 1022–1028.